Trauma and Mental Health in the Wake of a Technological Disaster

THE GHISLENGHIEN GAS EXPLOSION

Maklu
Antwerpen-Apeldoorn

Iedereen wil op de top van de berg leven zonder te weten dat,
het geluk zit in de manier waarop je de berg beklimt

Gabriel Garcia Marquez

Opgedragen aan mijn moeder en mijn overleden vader

Dit proefschrift is als handelseditie verkrijgbaar onder isbn 9789046607855

Maklu-Uitgevers
Somersstraat 13/15, B-2018 Antwerpen, www.maklu.be, info@maklu.be
Koninginnelaan 96, Nl-7315 EB Apeldoorn, www.maklu.nl, info@maklu.nl

Trauma and Mental Health in the Wake of a Technological Disaster
THE GHISLENGHIEN GAS EXPLOSION

Trauma en geestelijke gezondheid in de nasleep van een technologische ramp
DE GASONTPLOFFING VAN GELLINGEN
(met een samenvatting in het Nederlands)

Le trauma et la santé mentale au lendemain d'une catastrophe technologique
L'EXPLOSION DE GAZ DE GHISLENGHIEN
(avec un résumé en Français)

Proefschrift

ter verkrijging van de graad van doctor aan de Universiteit Utrecht op gezag van de rector magnificus, prof. dr. G.J. van der Zwaan, ingevolge het besluit van het college voor promoties in het openbaar te verdedigen op vrijdag 20 november 2015 des middags te 2.30 uur

door

Erik Leo Julien Louis De Soir
geboren op 4 november 1965
te Kinshasa, Kongo

Promotoren: Prof. dr. R.J. Kleber
Prof. dr. J. Mylle

CONTENTS

CHAPTER 1

Introduction

Introduction

Mental health professionals and researchers frequently face natural and human-caused disasters in their communities and may be called upon to assist the community and its residents following disasters or investigate the mental health consequences in the wake of these events.

In this thesis, mental health disturbances after the confrontation with a particular kind of technological disaster i.e. a massive gas explosion will be studied. Although each disaster has unique characteristics and dynamics, they all have in common the potential trauma of victims and disaster responders.

Explosions might happen due to an accident, a human error or a technical failure, but they may also be the consequence of a deliberate and intentional act aimed at injuring or killing innocent people in the context of terrorism. An explosion is always abrupt and shocking, and may involve disturbing feelings of anxiety and/or panic. By its sudden impact, such an event may temporarily disrupt the emotional, physical, cognitive and/or health equilibrium of the involved individuals and their significant others, being struck by the secondary impacts of the event. Survivors of an explosion may react in a passive (or even pathological) way, but they may as well show active coping responses based on social, and even helpful behaviour, fighting back against 'the evil' and recover without long term psychological sequelae.

The literature on massive explosions is very diverse, but their impact and consequences may have analogue consequences on survivors and rescue services personnel. This thesis describes the mental health consequences for the affected population of the Ghislenghien gas explosion in Belgium in 2004. This introductory chapter presents a general overview of the specificities of technological disasters in relation to the mental health disturbances they may generate. Several determinants of these mental health disturbances will be shortly discussed before introducing the Ghislenghien disaster study and the research questions dealt with in this thesis. The impact of that disaster will be studied through a phenomenological and an empirical analysis. The lived experiences of a miraculous survivor and testimonies of rescuers will lead to a better understanding of the psychological burden inflicted by a sudden and massive explosion. A theoretical interpretation of what this kind of event means for its survivors, is based on the current French (hermeneutical) trauma concept in comparison with Anglo-Saxon trauma theory. The empirical study is based on the determinants of mental health disturbances found in adult and child victims of this disaster.

Technological disasters

Man versus technology, technology versus man, is the central topic in technological disasters. Nature and politics kill far more people than technological accidents but failures of machines and infrastructure, and human error, negligence or incompetence still take an impressive toll on both a personal and social level.

Technological disasters can be categorized by the nature of their origin, by the effects they cause and by the responses of individuals and groups to the event. In general, they are dangerous, overwhelming and sudden, causing fear, anxiety, withdrawal, and avoidance.

Technology has a dual character. It is able to prevent disasters and to cause disasters. By definition (Ursano, Caughey, & Fullerton, 1994), technological disasters are the result of a failure of a human made product or a human error. They include airplane crashes, large scale road accidents, train derailments and collisions, passenger ship and other maritime catastrophes (including oil rig destruction), oil blowout, mining disasters, nuclear plant accidents, leakages of hazardous substances from toxic waste disposal, industrial explosions, etc. There is a lack of consensus in the literature about what constitutes a technological disaster. The heterogeneity within this category makes it difficult to draw general conclusions about the psychological effects of technological disasters (Myers & Wee, 2005). Despite the difficulties of categorizing disasters, some attempts have been made to differentiate psychological consequences based on the uniqueness of each technological disaster.

Baum, Fleming, and Davidson (1983) define technological catastrophes as events that are human-made in that they are accidents, failures, or mishaps involving the technology and manipulation of the environment that has been created to enhance the standard of living. They invariably cause resentment and blame among victims (Baum et al., 1983; Sorensen, Soderstrom, Coperhaver, Carnes, & Bolin, 1987). People usually view technology as something which should be under control. Therefore, after technological disaster, confidence in future controllability of technology is likely to be eroded (Baum et al., 1983). Some studies and authors claim that the more there is human causation behind a disaster, the more pathogenic it will be in terms of psychiatric morbidity (Baum, Fleming, Israel, & O'Keefe, 1992). Smith et al. (1988) report that rates of psychiatric disturbance following technological disaster vary as greatly as the number of disasters studies. Green, Grace, and Gleser (1985) caution, however, that these differences between studies in findings of disaster impact may be attributable to how, when, and on whom the data were collected. Smith et al. (1988) provide a thorough discussion of technological disasters and various factors that may influence their effects on human beings. In attempting to illuminate differences between disasters, they cite Solomon (1989) in summarizing that previously proposed differences in mental health effects of the different kinds of disaster might be explained on the basis of the bias in various contributors to the overall severity of the disaster agent rather than on the intrinsic basis of a possibly arbitrary division between the types of disasters themselves.

Drury, Novelli, and Stott (2013a, 2013b) studied the July 7th London bombings and underpin that survivors showed a sense of collective resilience in the way in which they expressed solidarity and cohesion through their active coping behaviour, supporting each other in order to adaptively deal with adversity. Although, it may be the specific context and the abruptness of events that cause mental health disturbances in survivors.

The absolute number of technological disasters is increasing. As technology develops, the quality of control and command systems may increase, but there are always things that can go wrong, even if unintended and uncalculated. In a technological disaster, a human action, or a product of human hand (a failed technology), results in the disruption of a community, and, at times, considerable death, injury and destruction.

Technological disasters overwhelm the emergency services, at least in the initial phase. According to Ursano et al. (1994), disaster workers may need help in the hours or days shortly after exposure. During this time, they often report high levels of discomfort, both physical and psychological. Direct exposure to death is a significant stressor that can may create mental health disturbances in rescuers. The rescue workers may be overwhelmed through the senses: viewing, smelling, touching, experiencing the grotesque, the unusual, the novel and the untimeliness of death. Therefore, it is necessary that disaster research should not only focus on the post-disaster physical and/or mental health consequences in victims and their relatives, but also on the impact of the numerous caregivers and first responders involved in the search and rescue interventions.

The survivors of a technological disaster may be confronted with extreme life-threat and/or the painful or sudden death of a friend or a relative, seeing severely injured or dead people and other important losses. When people witness severely wounded victims after accidents or fires they may be psychologically disturbed and temporarily disrupted from their daily life routines but not necessarily traumatised.

The experiences of survivors of industrial or technological accidents, wars, fires and explosions are potentially life-changing. From then on, their lifespan is divided in 'before the disaster' and 'after the disaster'. The overwhelming forces annihilate human values, norms and/or life. Impressive amounts of violence and power contain the potential to reduce the human being to dust, extinguished life, leaving the survivors in the sometimes extremely short but impressive silence of emptiness, complete abandonment and loneliness, typical to the immediate aftermath of technological disaster.

Mental health disturbances after technological disaster

Ursano, Caughey, and Fullerton (1994) stated that in Western countries, 5% of all deaths are due to the consequences of aggressive or unnatural, technological, events, but it is unclear how this percentage evolved some twenty years later. Major disasters and accidents belong to this category of events.

The intrusive recollections and re-experiences (very strong memories, images or nightmares with reference to the traumatic event), during which the survivors return to the hyperaroused states of the original *fight to survive* (strong startle responses, heightened alertness and sleep disturbances), alternated by moments (or periods) of denial, avoidance (of thoughts, places or people linked to the traumatic event), social disruption and isolation, and alterations of mood and basic cognitions about the world, are the signature of what is described in the literature as *posttraumatic stress disorder (PTSD)* in the *Diagnostic and Statistical Manual for Mental Disorders (DSM)-IV* (APA, 1994) and DSM-5 (APA, 2013). When these symptoms cause such a severe suffering that it hampers efficient social, familial or professional functioning for more than one month, they are indicative of PTSD.

Earlier studies document that PTSD is present in 1 to 11 % of the entire population during the first year after a disaster (Heinrichs et al., 2005; Foa, Keane & Friedman, 2000; Kessler, Sonnega, Bromet, Hughes & Nelson, 1995; Ursano et al., 1994). A higher prevalence is shown for people directly involved in a disaster (25 to 75 %) (Galea, Nandi & Vlahov, 2005).

Up to now, no Belgian study had been conducted into the impact of a technological disaster on public health or into the prevalence of posttraumatic stress disorder (PTSD) following technological disasters. To our knowledge, only one study on PTSD (Maes et al., 1998) investigated the impact of respectively two major accidents: 1) a sudden fire blaze in a ballroom in which about 450 guests were celebrating New Year Eve 1994/1995 (Switel Hotel, Antwerp, Belgium), in which 15 people died on scene or later, more than 160 subjects were injured or burned; and, 2) a massive car pile-up (on 1996 February 27th) involving about 200 cars and trucks, due to a sudden heavy fog bank (on the E-17 highway at Nazareth, Belgium), in which 10 people died on the scene of the accident and 50 others were injured. The incidence of PTSD for these qualitatively different accidents appeared to be 26.2% for the ballroom fire blaze and 16,4´% for the massive car pile-up.

The varying degrees of PTSD in disaster survivors show that scientific research is not univocal with regard to the determinants of PTSD. Many studies show methodological shortcomings which might be inherent to the characteristics of research on the impact of disasters on (mental) health : the suddenness of a disaster, the lack of control group and thus uncontrolled designs (Ursano & Fullerton, 1997), completely randomized designs, the different definitions for the concepts used (regarding the type of traumatic event, psychological trauma, dissociation, etc.) or for the victim categories (Brewin, Andrews & Valentine, 2000; Hoogduin, Van Minnen, Verbraak & Van de Griendt, 2002; Miller, 2003; Raphael & Wilson, 1993; Tierney, 2000). Therefore, it is important to investigate the determinants of post-disaster mental health disturbances in a longitudinal study design. We will focus on three risk factors in particular: peritraumatic dissociation, social support and degree of exposure.

Peritraumatic dissociation

In the past decade, scientific research has attributed a lot of importance to the study of risk factors concerning the development of psychological trauma. In particular, the predictive role of several phenomena occurring during or immediately after the traumatizing impact has been put forward. Reactions experienced immediately at the moment of the potentially traumatic event, such as peritraumatic dissociation, extreme anxiety, panic and negative emotions appear to be important predictors of persisting symptoms such as PTSD symptomatology (Bernat, Ronfeldt, Calhoun

& Arias, 1998). Among the various PTSD predictors, peritraumatic dissociation appears to be one of the most important factors.

According to Marmar, Weiss, Metzler and Delucchi (1996), peritraumatic dissociation consists of the appearance of dissociative symptoms during or immediately after the exposure to extreme events. Symptoms subsumed under the label of peritraumatic dissociation involve alterations of cognitive and perceptual functioning at the moment of, or directly after the traumatizing event. Traumatized individuals often report alterations in their experience of time, their sense of direction and self-esteem which leads to an impression of unreality with regard to the ongoing event (Marmar et al., 1996). Nijenhuis et al. (1998) suggested that peritraumatic dissociation is a manifestation of failed integration attempts with regard to the traumatic experience. Peritraumatic dissociation is linked to the perception of a larger life threat and to a larger loss of control (Marmar et al., 1996), to helplessness and anger (Simeon, Greenberg, Knutelska, Schmeidler & Hollander, 2003) and to hyperagitation (Sterlini & Bryant, 2002).

According to Schooler, Dougall and Baum (1999) the recollection of memories linked to potentially traumatizing experiences is valid and liable. In their comprehensive meta-analysis of predictors of posttraumatic stress symptoms and PTSD, assessed in studies published between 1980 and 2000, Ozer, Best, Lipsey and Weiss (2003) found that peritraumatic dissociation turned out to be the best PTSD predictor in comparison with other predictors such as earlier traumatization, earlier psychological well-being, familial antecedents of psychopathology, life threat felt during the trauma, social posttrauma support and emotional reactions occurring during or right after the event (peritraumatic reactions).

According to Van der Velden and Wittman (2008), a limitation in most studies is the absence of the evaluation of possible confounding variables in the relationship between peritraumatic dissociation and PTSD. For instance, controlling for strong emotional reactions and mental health problems occurring in the first days or weeks post-event, could modify findings about peritraumatic dissociation. This is in line with other findings, which suggest that the recollection of peritraumatic dissociation during or immediately after the particular event may be biased by the current psychological state of the affected individual (Bryant, 2007; Candel & Merkelbach, 2004; Marshall & Schell, 2002).

In their meta-regression which included 59 independent empirical studies, Lensvelt-Mulders et al. (2008) established a significant positive effect (r=.40) of peritraumatic dissociation on subsequent posttraumatic stress. Although this result is in line with the finding of Breh and Seidler (2007) who found a correlation of .36 in a meta-analysis of 34 studies, it does not suggest that experiencing dissociation during or shortly after a potentially traumatizing event increases the probability of posttraumatic stress later in life. Lensvelt-Mulders et al. (2008) argue that causality can only be proved using a rigorous prospective design. McDonald et al. (2013) provide evidence that peritraumatic dissociation is linked to the amount of time that trauma survivors perceive they had to anticipate the impact of the traumatic event.

To our knowledge, there are no studies on technological disasters in which the predictive value of peritraumatic dissociation with regard to PTSD has been evaluated. Therefore, the study of peritraumatic dissociation in the context of a sudden massive explosion, leaving no time for the victims to be prepared, is relevant for the discussion concerning the risk factors for development of PTSD symptoms in survivors of the disaster.

Social support

People tell others – mostly their intimates – about the emotional event they faced and about their related feelings, reactions and thoughts (Rimé, Mesquita, Philippot, & Boca, 1991) but what is the effect of talking about potentially traumatizing experiences? Such a sharing process occurs in 80 to 95% of all emotional episodes and usually develops in the period immediately following an emotional event (Rimé, Philippot, Boca, & Mesquita 1992; Rimé, Finkenauer, Luminet, Zech, & Philippot, 1998). Modally, the sharing of an emotion is repetitive and addresses several successive recipients. Experimental studies confirmed emotional exposure to cause the social sharing of emotion (Luminet, Bouts, Delie, Manstead, & Rimé, 2000).

Emotion sharing was consistently found to hold a positive linear relationship with the intensity of the emotional experience (Rimé et al., 1998). Thus, more intense episodes are shared more repetitively and with a superior number of successive targets. This positive relationship leads to expectation that emotions of extreme intensity and potentially traumatizing experiences would elicit particularly abundant manifestations of emotion sharing. This was corroborated by Janoff-Bulman (1992) who stressed that people who went through a traumatic experience later evidence a seemingly insatiable need to tell others about it, as if they felt coerced into talking. Empirical data in this respect were surprisingly scarce. A need to talk was mentioned by 88 % of rescuers operating in a North Sea oil platform disaster (Ersland, Weisaeth & Sund, 1989) and by 88 % of people who had recently lost a relative (Schoenberg, Carr, Peretz, Kutscher, & Cherico, 1975). Similarly, in a large sample of Belgian expats who lived through the Rwandan genocide of 1994, 98% reported that they shared this experience at least once, and the modal response (64%) was 5 to 6 times (Sydor & Philippot, 1996). Three to five months later, most of them (90%) reported that they were still sharing these events, with 36% of them still sharing frequently or very frequently.

Most available data support the linear relationship between intensity of emotion and extent of social sharing. Victims of extreme emotional experiences do seem to share their emotion as frequently as individuals who experienced a highly emotional daily life situation. However, taking into account the diagnostic criteria for PTSD (DSM-IV, American Psychiatric Association, 1994) would lead to predict variations in this regard as a function of whether people exposed to a potentially traumatizing event developed, or not, this disorder. Indeed, a major diagnostic criterion of PTSD, criterion C, includes avoidance of cues and memories associated with the traumatic experience. Consistent with this criterion, one would expect those who were exposed to a potentially traumatic event and later met the diagnostic criteria for PTSD to be much less prone to socially share their experience than those who were exposed to a comparable situation, but remained free of PTSD (free of symptoms or free of the full diagnostic PTSD image). Whether the former would predominantly remain silent, or would refuse to talk, or would manifest a partial concealment of their experience could hardly be specified in absence of empirical observations.

Since disasters always strike groups of people, it is also important to investigate the effect of social support. Previous research (Brewin, Andrews, & Valentine, 2000; Guay, Billette, & Marchand, 2006: Ozer et al., 2003) found that social support counters the development and maintenance of distress and PTSD.

Degree of exposure

The posttraumatic consequences (in terms of prevalence) also depend on the type of exposure to the disaster ; for instance, a PTSD prevalence of 35 % two months after accidents with serious injuries (Perry, Dean & Krenzelok, 1992) ; 7 to 20% five to eight weeks after the disaster depending on the proximity to the World Trade Center site (Galea, Resnick, Ahem, Gold, Bucuvalas & Kilpatrick, 2002) ; and, 25 % ten years after a disaster on an oil platform (Hull, Alexander & Klein, 2002). Determining the degree and type of exposure to a technological disaster is complex. An epidemiologic perspective is critical to identifying the at risk population and the probability of post disaster mental health disturbances. The comparison between different disasters is not always meaningful, since the timing of the measurements of PTSD symptoms and the used instruments are too diverse. To our knowledge, there is no previous study which compares the mental health disturbances between survivors of the same disaster as a function of their exposure during the disaster.

Various kinds of victims may be affected by a disaster. Bolin (1986) describes the importance of victim characteristics in disaster vulnerability. He identifies individuals as *primary victims* (those directly experiencing physical, material, and personal losses) and *secondary victims* (those who witness the disaster but do not experience the actual impact). While both levels of victims may have comparable levels of psychological distress, he points out that secondary victims are less likely to seek professional mental help. Next to the primary and secondary victims, also rescue workers, mental health workers and witnesses may suffer from the consequences of the disaster according their degree of exposure; they are the *tertiary victims.*

Groups with differing degrees of exposure to a disaster can provide a 'dose' variable to study the response effects to the disaster. Control groups, i.e. individuals who were less exposed or unexposed to the potentially traumatizing effects of the disaster and matched with the direct victims on certain characteristics, allow researchers to examine the consequences as a function of the degree of exposure.

DeWolfe (2000) also describes a direct dose-response relationship between community impact and psychological impact. When entire communities are destroyed and familiar infrastructure is destroyed, survivors become more disoriented at the most basic levels. Researchers have found higher levels of anxiety, depression, posttraumatic stress, somatic symptoms, and generalized distress associated with widespread community destruction (Solomon & Green, 1992). Palinkas, Russel, Downs, and Petterson (1992) found that 43% of people highly exposed to the environmental damage had one or more psychological problems, compared to 23% of those not exposed. Severity of exposure also predicted declines in social relations and increased conflicts with family members.

The Ghislenghien disaster

On July 30th 2004, an accidental leakage in a high pressure gas pipe, which passed under the industrial zone of Ghislenghien (Belgium) created a persistent gas smell and alerted the employees of one of the factories. When the first crew of firefighters arrived on-scene, an enormous explosion took place and instantly killed 24 people (including five firefighters and one police officer). From the first crew, only two firefighters survived the initial blast and 132 people were wounded. An impressive column of fire rose into the air and debris from the gas pipe was thrown up high into the air. The heat of the fire was felt up to two kilometers away from the explosion site. Debris from buildings was projected up to six kilometers away from the epicentre and up to 16 km from the explosion, air vibrations were registered. Hundreds of fire, rescue and police personnel rushed to the disaster area and all the regional hospitals received numerous victims. A wide area was affected by the largest technological disaster that Belgium ever knew.

The Ghislenghien disaster happened on a beautiful and sunny day; it was in the morning of July 30th, 2004 at 8:54 am, the outside temperature at the industrial area was 17.5°C. The survivors remember this "hell" that destroyed entire lives and provoked a severe trauma for a lot of victims. The experience on "Ground Zero" must have been comparable to what victims of massive terrorist attacks and war experience.

The detonation stunned everyone who was present on the disaster site. In their accounts, survivors report the tremendous noise which they did not understand. Many of them were thinking of an airplane crash. Some were even thinking about a nuclear explosion when they saw the mushroom shape of the enormous flame that rose until 500 meters in the air. Soon, the temperature at the disaster scene went up to more than 300°C urging the survivors to flee the area like animals under the direct threat of a predator. They will always remember the screams of others, the sights of horror and the odour of burned keratin.

Fire, rescue and police services rushed to the area and the medical disaster plan was activated. The victims were evacuated into a school used as hospital of misfortune where the medical triage and regulation to hospitals and burn centers in Belgium and The Netherlands went on for the rest of the day. The degree of exposure for direct involved victims and their caregivers must have been enormous: the initial chaos, the cries, the smell and the deformed bodies. Enough to lose each sense of reality. An incredible force of technology on human being. Families needed to be alerted and an extensive framework of psychosocial assistance was put in place. As direct victims were being helped, the indirect victims received the corollary shock, the retroactive effect of receiving the bad news about deceased or severely wounded family members of friends. The 'disaster after the disaster' consisted of the horrifying experiences of burn injuries in the hospitals transforming the initial euphoria of survivors into despair.

Hades, the Greek God of Hell, seemed to be after the victims that day. In a hell of fire and heat, they were confronted with the apotheosis of inhumanity: victims burning to death in a suffocating heat, feeling abandoned and in the hands of God.

Mental health disturbances and an extensive negative impact on general health in survivors and their family members were to be expected. For that reason, it seemed essential to implement a large study assessing these consequences among the population of the whole region affected by one of the largest technological disasters that Belgium even knew.

On the July 30th, 2004, it was my French neighbor on a camping in the Southern part of France that alerted me. I switched on my television and saw the horrible consequences of the Ghislenghien disaster. My telephone started to ring around 10am in the morning and would not stop ringing for the coming days; friends from the regional fire services, colleagues psychologists, the media, etc. I decided to stop my vacation and returned back to Belgium.

In 1993, I created the Fire Stress Teams and since five firefighters died in the explosion, our teams got involved immediately together with the psychologists of the Service d'Appui Psychologique (Psychological Support Service) from the Institut Provincial de Formation lead by Frédéric Daubechies. Through the psychological debriefing sessions with shocked firefighters, police services, emergency medical services personnel, local authorities and psychosocial caregivers, I also got into direct contact with the victims and their families. I started as a crisis psychologist in the front line, but would stay involved in the psychosocial aftercare for years. Together with Emmanuelle Zech, a colleague from the Université Catholique de Louvain, we decided to create a multidisciplinary research team and convinced the authorities to initiate a series of longitudinal scientific studies which became the building bricks of this thesis.

The Ghislenghien disaster studies

The Ghislenghien disaster study on general health

On our request, the Belgian Federal Public Service Health, Food Chain Safety and Environment initiated a longitudinal study in order to evaluate the overall impact of the disaster on the general health of the affected population, with a special focus on mental health. This study aimed at assessing the general and mental health consequences of this technological disaster and improving disaster aid structures in Belgium. The current study was carried out by the Institute of Public Health Unit of Epidemiology (Brussels) in collaboration with the Faculty of Clinical Psychology of the Catholic University of Louvain and the Stress and Trauma Research Centre of the Royal Military Academy.

The Ghislenghien disaster study in fire, rescue and emergency medical services

In order to evaluate the psychological impact of the Ghislenghien disaster on the fire, rescue and emergency medical services personnel, I initiated another study under the lead of the Stress and Trauma Research Center of the Royal Military Academy, the Faculty of Psychology of the Université Cathololique de Louvain, the Service d'Appui Psychologique aux Intervenants (Psychological Support Service for Caregivers), and the Union Royale des Sapeurs-Pompiers du Hainant (Royal Association of Firefighters of the Province of Hainaut).

Fire, rescue, emergency services and in-hospital personnel were contacted through their normal command or management structures in order to participate in this longitudinal study and respond to two self-report questionnaires, respectively four and fourteen months after the disaster.

This study aimed at analysing the similarities and differences between the disaster-related experiences of firefighters and emergency medical personnel (including in-hospital rescue staff). A qualitative analysis method of the answers to open-ended questions was used to assess which disaster-related aspects were viewed as the most shocking.

Outline and research questions of this thesis

In this thesis, papers with different research questions on different survivors are presented in order to assess the mental health disturbances after the Ghislenghien disaster. The aim of this thesis is fourfold:

1. Highlight and discuss the theoretical differences between the concept of psychological trauma used in mainstream literature and the more hermeneutic trauma interpretation in French theory.
2. Evaluate the indication for PTSD and related reactions in survivors of a technological disaster (both in adults and children).
3. Investigate the mental health disturbances after a technological disaster originated by an accidental massive gas explosion.
4. Evaluate the prevalence of PTSD in survivors of a technological disaster (both in adults and children).

In general, the aim is to study and clarify the impact of a technological disaster, both phenomenologically and empirically in terms of mental health disturbances in the affected population. Therefore, the corpus of this thesis has been divided in two major parts: a phenomenological part and an empirical part.

Part I: Phenomenological part

This part investigates the lived experiences of disaster survivors and their rescuers:

Chapter 2 presents the personal and unique account of a survivor who qualified his survival as a traumatic journey with Hades. The chronology of the first moments is described as a succession of ten traumatic shocks and is explained relying on different theoretical trauma models. This chapter has been published in French in the *Revue Francophone du Stress et du Trauma* and raises the next questions:

- What is the lived experience from a survivor who directly witnessed the massive gas explosion at Ghislenghien?
- How can the lived experienced be explained through different theoretical approaches in psychotraumatology (hermeneutical vs empirical theories)?

Chapter 3 examines the conceptualization of psychological trauma in contemporary French theory and the clinical importance this approach still holds beyond the framework of Anglo-Saxon and/or English-speaking trauma clinicians. This chapter illustrates how a traumatic encounter with disaster events fits the French clinical description of psychological trauma based on the concept of *effroi de la mort* (terror of death) and the *syndrome de répétition* (repetition syndrome). It aims to clarify the conceptual differences between the mainstream theories, currently used by Anglo-Saxon researchers and practitioners, and the more classical psychodynamic theories which are still the basis for clinical practice in Latin countries (France, Spain, Italy, etc.). Living and waking on the border area between Northern Europe and France, for more than ten years, I have been the Vice-President of the *Association de Langue Française pour l'Etude du Stress et du Traumatisme (ALFEST)*, trying to introduce the clinically relevant French trauma concepts in the international community of trauma practitioners and researchers. As the Co-Chair (together with Danny Kaloupek) of the *International Structure & Affiliations Committee* of the *International Society for Traumatic Stress Studies*, I have contributed to the dissemination of these concepts on various conferences, but always the question was raised where this current French interpretation had been published. Unfortunately, these French concepts have only been published in French scientific journals and/or books. Therefore, an aim of this thesis is to bridge these conceptual differences and present the contemporary French trauma theory starting from the following questions:

- How can a better understanding of the French clinical description of psychological trauma offer a more integrative view on psychological trauma of trauma survivors?
- Could there be a contribution from this hermeneutical trauma concept to the current conceptual discussion on PTSD in DSM-5 and the upcoming ICD-11?

Chapter 4 explores the experiences of fire and emergency medical services personnel during and immediately after a technological disaster using a phenomenological approach. These experiences are described in the responses to the open-ended questions of personnel engaged in the rescue operations during and immediately after the Ghislenghien gas explosion and were qualitatively analysed. The lived experiences of fire-fighters, emergency medical personnel and in-hospital staff were compared. This chapter has been published (in two different versions) in *Prehospital and Disaster Medicine* and in the *Revue Francophone du Stress et du Trauma*. It aims to understand the differences in the impact of potentially traumatizing events on rescuers with a different role.

- With regard to potentially traumatizing events, what are the differences concerning the central characteristics of these events in different categories of rescuers? And, what are the differences regarding the most prominent emotions?
- What kind of difficult work-related experiences do fire-fighters report compared to emergency medical services personnel and in-hospital staff?

Part II: Empirical part

This part of the thesis consists of four chapters.

Chapter 5 describes the strengths and weaknesses of the methodology used for the Ghislenghien study design and discusses the issues concerning the overall response rate. This chapter aims to clarify the importance of the study for the psychosocial management of disasters.

Chapter 6 investigates the risk factors[1] for the development of posttraumatic stress symptoms in the different adult survivor groups involved in the Ghislenghien disaster. The research questions related to Chapter 6 are consecutively:

- What is the prevalence of posttraumatic stress symptoms in adult survivors of a gas explosion, in their family members as well in family members of deceased victims?
- Which are the predictors of the development and chronification of posttraumatic stress symptoms in the different survivor groups?
- What are the problems inherent to disaster research and how can they be addressed through changes in study design?

1 In current literature, terminology is sometimes confusing, mixing different but apparently similar concepts such as predictors, determinants, and, risk factors regarding the development of posttraumatic stress symptoms. In this thesis, I made the choice to use 'risk factor' as a common denominator for factors containing a predictive quality towards the development of posttraumatic symptoms.

Chapter 7 concerns the risk for the development of four types of mental health disturbances (MHD) due to exposure to the different aspects of this technological disaster in comparison with data obtained from previous health surveys among the population of the same province. This paper calls attention to MHD as a function of survivors who have been directly exposed to human damage during or after the gas explosion. The following research questions are addressed in this chapter;:

- What is the prevalence rate of MHD in a population affected by the disaster depending on the degree of exposure?
- How do these prevalences relate to those of a reference population of the same province.

Chapters 5, 6 and 7 have been published in *Archives for Public Health*

Chapter 8 describes the risk factors for the development of posttraumatic stress reactions in children after a technological disaster. This chapter illustrates how children may recover differently than adults as a function of the type of exposure, the severity of dissociation and the type, timing and perceived benefits of psychological help. It focuses on the following research question:

- How do children involved in technological disaster react and what are predictors risk factors for the development and chronification of their symptoms?

This chapter has been published in the *Journal of Child and Adolescent Trauma.*

In the margin of the above research questions, this thesis also aims to better understand the real challenges of the type and timing of psychological help in the wake of technological disaster.

The **Discussion and Conclusions** chapter proposes a set of possible answers to the research questions and contextualises the insights gained and lessons learned through the Ghislenghien disaster studies.

In the **Epilogue**, a model of psycho-physiological stabilization of severely wounded and/or shocked victims of motor vehicle accidents is introduced as a possible future good practice for acute trauma support of disaster survivors, starting on the scene of the accident.

References

American Psychiatric Association. Diagnostic and Statistical Manual for Mental Disorders, 4th EditionWashington DC: Author, 1994.

Baum, A., Fleming, R., & Davidson, L.M. (1983). Natural disaster and technological catastrophe. *Environment and Behaviour, 15,* 333-354.

Baum, A., Fleming, R., Israel, A., &, O'Keefe, M. (1992). Symptoms of chronic stress following a natural disaster and recovery of a human-made hazard. *Environment and Behavior, 24,*347-365.

Bernat, J., Ronfeldt, H., Calhoun, K., &, Arias, I. (1998). Prevalence of traumatic events and peritraumatic predictors of posttraumatic symptoms in a non-clinical sample of college students. *Journal of Traumatic Stress, 11(4),*645-664.

Brewin, C., Andrews, B., & Valentine, J. (2000). Meta-analysis of risk factors for posttraumatic stress disorder in trauma-exposed adults. *Journal of Consulting and Clinical Psychology, 68(5),*748-766.

Breh, D., & Seidler, G. (2007). Is peritraumtic dissociation a risk factor for PTSD. *Journal of Trauma and Dissociation, 8(1),*23-69. doi:10.1300/J229v08n01_04.

Bryant, R. (2007). Does dissociation further our understanding of PTSD? *Journal of Anxiety Disorders, 21,*183-191. doi:10.1016/j.janxdis.2006.09.012.

Candel, L., &, Merkelbach, H. (2004). Peritraumatic dissociation as a predictor of posttraumatic stress disorder: A critical review. *Comprehensive Psychiatry, 45,*44-50.

De Wolfe, D. (2000). Training manual for mental health and human service workers in major disasters (2nd edition). (DHHS Publication N°. ADM 90-538). Available at http://www.mentalhealth.org/publications/allpubs/ADM90-538/index.htm

Dougall, A., Craig, K., & Baum, A. (1999). Assessment of characteristics of intrusive thoughts and their impact on distress among victims of traumatic events. *Psychosomatic Medicine, 61,*38-48.

Drury, J., Novelli, D., & Stott, C. (2013a). Representing crowd behaviour in emergency planning guidance: 'Mass panic' or collective resilience? *Resilience: International Policies, Practices and Discourses, 1*(1), 18–37. doi.org/10.1080/21693293.2013.765740

Drury, J., Novelli, D., & Stott, C. (2013b). Psychological disaster myths in the perception and management of mass emergencies. *Journal of Applied Social Psychology, 43,* 2259–2270. doi: 10.1111/jasp.12176.

Ersland, S., Weisaeth, L., & Sund, A. (1989). The stress upon rescuers involved in an oil rig disaster: "Alexander L. Kielland" 1980. *Acta Psychiatrica Scandinavia, 80 (Suppl.355),* 38-49.

Foa, E., Keane, T., & Friedman, M. (2000). Guidelines for treatment of PTSD. *Journal of Traumatic Stress, 13(4),*539-588.

Heinrichs, M., Wagner, D., Schoch, W., Soravia, L., Hellhammer, D., & Ehlert, U. (2005). Predicting posttraumatic stress symptoms from pretraumatic risk factors: A 2-year prospective follow-up study in firefighters. *American Journal of Psychiatry,162,*2276-2286.

Hoogduin, C., Van Minnen, A., Verbraak, M., & Van de Griendt, J. (2002). Klinisch beeld van de PTSS. In: A. Van Minnen & M. Verbraak (Eds.), *Psychologische interventies bij posttraumatische stress.* Nijmegen: Cure & Care Publishers, 1-14.

Galea, S., Resnick, H., Ahern, J., Gold, J., Bucuvalas, M., &, Kilpatrick, D. (2002). Posttraumatic stress disorder in Manhattan, New York City, after the September 11th terrorist attack. *Psychological Science, 17(3),*181-186.

Galea, S., Nandi, A., & Vlahov, D. (2005). The epidemiology of post-traumatic stress disorder after disasters. *Epidemiological Review, 27,*78-91.

Green, B., Grace, M., &, Gleser, G. (1985). Identifying survivors at risk: Long term impairment following the Beverly Hill Supper Club Fire. *Journal of Consulting and Clinical Psychology, 53,*672-678.

Guay, S., Billette, V., &, Marchand, A. (2006). Exploring the links between posttraumatic stress disorder and social support: Processes and potential research avenues. *Journal of Traumatic Stress, 19(3),*327-338.

Hull, A., Alexander, D., &, Klein, S. (2002). Survivors of the Piper Alpha oil platform disaster: Long-term follow-up study. *Britisch Journal of Psychiatry, 181,*433-438.

Janoff-Bulman, R. (1992). *Shattered assumptions: Towards a new psychology of trauma.* New York: Free Press.

Kessler, R., Sonnega, A., Bromet, E., Hugher, M, & Nelson, C. (1995). Posttraumatic stres disorder in the national comorbidity survey. *Archives of General Psychiatry, 52(12),* 1048-1060.

Lensvelt-Mulders, G., Van der Hart, O., Van Ochten, J., Steele, K., & Breeman, L. (2008). Relations among peritraumatic dissociation and posttraumatic stress: A meta-analysis. *Clinical Psychology Review, 28,*1138-1151.

Luminet, O., Bouts, P., Delie, F., Mainstead, A., &, Rimé, B. (2000). Social sharing of an emotion following exposure to a negatively valenced situation. *Cognition & Emotion, 14,* 661-668.

Maes, M., Delmeire, L., Schotte, C., Janca, A., Creten, T., Mylle, J. Struyf, A., Pison, G., Rousseeuw, P. (1998). The two-factorial symptom structure of post-traumatic stress disorder: depression-avoidance and arousal-anxiety. *Psychological Research, 81,* 195-210.

Marmar, C., Weiss, D., Metzler, T., & Delucchi, K. (1996). Characteristics of emergency services personnel related to peritraumatic dissociation during critical incident exposure. *American Journal of Psychiatry, 153(7 Suppl),*94-102.

Marshall, G., &, Schell, T. (2002). Reappraising the link between peritraumatic dissociation and PTSD symptom severity: Evidence from a longitudinal study of community violence survivors. *Journal of Abnormal Psychology, 111*,626-636.

McDonald, P., Bryant, R., Silove, D., Creamer, M., O'Donnell, & McFarlane, A. (2013). The expectancy of threat and peritraumatic dissociation. *European Journal of Psychotraumatology, 4*. doi: 10.3402/ejpt.v4i0.21426.

Miller, M. (2003). Personality and the etiology and expression of PTSD: a three factor model perspective. *Clinical Psychology Review, 10(4)*,373-393.

Myers, D., & Wee, D.F. (2005). *Disaster mental health services*. New York & Hove: Brunner-Routledge.

Nijenhuis, E., Spinhoven, P., Vanderlinden, J., Van Dijck, R., &, Van der Hart, O. (1998). Somatoform dissociative symptoms as related to animal defence reactions to predatory imminence and injury. *Journal of Abnormal Psychology, 107*,225-258.

Ozer, E., Best, S., Lipsey, T., &, Weiss, D. (2003). Predictors of posttraumatic stress disorder and symptoms in adults: A meta-analysis. *Psychological Bulletin, 129(1)*,52-73.

Palinkas, L., Russel, K., Downs, M., & Petterson, J. (1992). Ethnic differences in stress, coping and depressive symptoms after the Exxon Valdez oil spill. *Journal of Nervous and Mental Disorders,180*,287-295.

Perry, P., Dean, B., & Krenzelok, E. (1992). A regional poison center's experience with poisoning exposure occuring in schools. *Veterinary and Human Toxicology, 34(2)*,148-151.

Raphael; B., & Wilson, J. (1993). Theoretical and intervention considerations in working with victims of disaster. In: B. Raphael, & J. Wilson (eds.). *International handbook of traumatic stress syndromes*. New York: New York Plenum, 105-117.

Rimé, B., Mesquita, B., Philippot, p., & Boca, S. (1991). Beyond the emotional event – 6 studies on the social sharing of emotion. *Cognition & Emotion, 5(5-6)*,435-465.

Rimé, B., Philippot, O., Boca, S., & Mesquita, B. (1992). Long lasting cognitive and social consequences of emotion : Social sharing and rumination. *European Review of Social Psychology*, 3, 225-258.

Rimé, B., Finkenauer, C., Luminet, O., Zech, E., & Philippot, P. (1992). Long-lasting cognitive and social consequences of emotion: social sharing and rumination. In: M. Stroebe & M. Hewstone (Eds.) *European Review of Social Psychology* (pp.225-258). Chichester: Wiley.

Schoenberg, B., Carr, A.C., Peretz, D., Kutscher, A.H., & Cherico, D.J. (1975). Advice of the bereaved for the bereaved. In B. Schoenberg, I. Gerber, A. Wiener, A.H. Kutscher, D. Peretz & A.C. Carr (Eds), *Bereavement. Its psychosocial aspects* (pp. 362-367). New York: Columbia University Press.

Simeon, D., Greenberg, J., Knutselska, M., Schmeidler, J., & Hollander, E. (2003). Peritraumatic reactions associated with the World Trade Center disaster. *American Journal of Psychiatry, 160(9)*,1702-1705.

Smith, E., North, C., & Price, P. (1988). Response to technological accidents. In M.Lystad (Ed.), *Mental health response to mass emergencies: Theory and practice.* New York: Brunner/Mazel.

Solomon, S., Smith, E., Robins, L., & Fischbach, R. (1987). Social involvement as a mediator of disaster-induced stress. *Journal of Applied Psychology, 17*,1092-1112.

Solomon, S. (1989). Evaluation and research issues in assessing disaster's effects. In R.Gist & B.Lubin (Eds.), *Psychosocial aspects of disaster.* New York:Wiley.

Sorensen; J., Soderstrom, J., Coperhaver, E., Carnes, S., & Bol R. (1987). *Impacts of hazardous technology: the psycho-social effects of restarting TMI-1.* Albany, New York: State University of New York Press.

Sterlini, G., & Bryant, R. (2002). Hyperarousal and dissociation: a study of novice skydivers. *Behaviour Research Therapy , 40(4)*,431-437.

Sydor, G., & Philippot, P. (1996). Prévalence des symptômes de stress post-traumatique et intervention de prévention secondaire suite à une catastrophe humanitaire. *Revue Européenne de Psychologie Appliquée, 46*, 269-275.

Tierney, K. (2000). Controversy and consensus in disaster mental health research. *Journal of Prehospital and Disaster Medicine, 15(4)*,181-187.

Ursano, R., & Fullerton, C. (1997). *Trauma, time and recovery. Posttraumatic stress disorder. Acute and long-term responses to trauma and disaster.* Washington, DC: American Psychiatric Press.

Ursano, *R., Caughey, B., & Fullerton, C. (1994). Individual and community responses to trauma and disaster: The structure of human chaos.*Cambridge: Cambridge University Press.

Van der Velden, P., Witman, L. (2008). The independent predictive value of peritraumatic dissociation for PTSD symptomatology after type 1 trauma: A systematic review of prospective studies. *Clinical Psychology Review, 28(6)*,1009-1020.

PART I

The Phenomenology of Psychological Trauma in a Technological Disaster

CHAPTER 2

My Traumatic Journey with Hades: Surviving a Massive Explosion in Ten Progressive Shocks

The Lived Experience of a Survivor of the Ghislenghien Disaster

E. De Soir & J. Mylle

Based on French paper published in *Revue Francophone du Stress et du Trauma* – Juin 2010 – Tome 10 – N°2

My Traumatic Journey with Hades: Surviving a Massive Explosion in Ten Progressive Shocks

The Lived Experience of a Survivor of the Ghislenghien Disaster

Erik De Soir & Jacques Mylle

Summary

This article describes the experiences of a survivor of the Ghislenghien disaster, by means of his testimony regarding the *progressive traumatization* of his survival. The chronology of the first moments is described as a traumatic journey to Hades and is explained relying on different models; i.e. the current empirical trauma models, on the one hand, and the psychodynamic theories developed by influential contemporary French authors, on the other hand. Particularly the moments of peritraumatic dissociation – i.e. those of derealization, of acting on automatic pilot and of reality denial – appear to have had an inerasable impact. The experience of frozen fright, petrification and horror, has left this survivor with a traumatic impact which developed gradually and proportionally with his pain as a severely burnt victim and miraculous survivor.

This testimony makes it possible to uncover the core of traumatic survival and identify essential factors in the primary and secondary prevention of posttraumatic sequelae – some of which could be implemented by field rescue workers who should understand the meaning but also the pitfalls of the peritraumatic defense mechanisms acting on acutely traumatized victims.

Key words: technological disasters, traumatogenic events, peritraumatic dissociation, neurobiological models, psychodynamic theories

Introduction

The primary goal of this article is to describe the experiences of S.D., based on a long individual clinical session concerning his miraculous survival of the technological disaster in Ghislenghien. The lived experience of this survivor may lead to a better understanding of the peritraumatic and posttraumatic experiences of victims of a massive and unexpected explosion. The experiences of survivors of this kind of sudden explosions are of utmost importance to understand the wartime experiences of bombing victims and/or the support to people who can be confronted with similar detonations.

A deep-level analysis of these experiences will allow for a confrontation between different trauma theories; more specifically, the essentially empirical and neurobiological trauma interpretations, on the one hand, and the more classic and contemporary French trauma theories, showing rather psycho-analytic or psychodynamic interpretations, on the other hand. This confrontation of concepts and interpretations will show that the older French authors, such as Pierre Janet, who started from their own clinical observations, knew already remarkably well what seems to happen in the mind of the traumatized victim.

Secondly, it is important to show that advanced psychological support may prevent the further traumatization of disaster victims at the secondary level. This kind of 'field practice' could possibly involve providing immediate psychological support, if possible in the peritraumatic phase, even if it has not found its legitimacy yet in the mind of emergency medical personnel and/or health professionals, and even if it continues to give rise to numerous questions. It is possible that disaster victims develop a secondary traumatization during and after the rescue operations and/or consecutively to their experiences, on site of the explosion as well as during their transportation to and/or the arrival at the hospital.

Testimony of a survivor: my journey with Hades

In the following paragraphs, S.D. testifies (cf. italic text) about the different stages of his survival, in ten chronological and progressive shocks.

The Ghislenghien disaster has upset hundreds of thousands of people around the world because the disaster caused twenty-four deaths and one hundred and thirty-two casualties; it was in the morning of July 30th, 2004 at 8:54 am, the outside temperature at the industrial area was 17.5°C. Some people remember this "hell" that destroyed entire lives and provoked a severe trauma for a lot of victims. I belonged to the latter! Indeed, at the moment of the disaster, I was in one of the buildings at the opposite side of the detonation. One of my responsibilities was to check the number of working hours of the job students. That is why, about ten minutes before the explosion, I asked a student who finished his job on that particular day to come, but he would never survive the catastrophe...

On that July 30th, 2004, I should have died three times: 1) a first time, when working in my car while listening to the radio; after some hesitation I decided to work in my office in order to send my work straight to Brussels by fax, notwithstanding the tremendous noise of the nearby building yard; 2) by joining a deceased on the place of the explosion; one of the leaders of the building yard found himself on "Ground Zero", after having finished my first task, I had to go round there to take note of the situation...; and finally, 3) by being at the mercy of the mud.

The first shock: start of the journey to hell

At the moment of the detonation, there were three of us in the room: the daughter of a victim who passed away while searching for her (student worker for July), another friend severely injured and myself. Right after a tremendous noise I did not try to understand, I had to get up and leave without asking myself what was happening, like many of us... This was the first shock for me; as I ran away, I bumped into a victim who would pass away that same day while searching for his daughter. The deceased looked into my eyes as if to ask me "and my daughter!?" As far as my look is concerned, it seemed to say "Yvan, what is happening?" The redness from outside to the way out as well as the dust lifted up by the blast of the explosion felt like true hell! As I knew the building very well, I decided to flee to the backside of the factory, without knowing what was happening, without understanding the complete disaster...

The second shock: the thermal shock, unconceivable and bewildering

When I arrived at the left wing of the building through the emergency exit, I was immediately thunderstruck by the thermal shock! I remind you that the outside temperature in the morning was 17.5°C! We switched from 17.5°C to more than 300°C in the blink of an eye, which drove us mad immediately! Usually, when we burn ourselves, our mind has well internalized the danger flames or a burning recipient can pose to our body but that doesn't make the burns less painful... We have to realize here at this stage that the enemy remains unknown; and precisely this drives anyone mad.

The third shock: the feeling of powerlessness and desolation

Already at this stage, I had arrived in hell! It was completely unknown; I did not recognize the place where I came from an hour and a half ago... Even the idea of a nuclear explosion crossed my mind, because this horror was so unknown to me, the heat was so terrible that it threw me to the ground, the sky seemed so red, we all screamed so hard! The unknown traumatized me ...

Fleeing seemed to be the only way out but when I slipped onto the ground, I realized how cool the mud felt, a clayey soil. Instinctively, without thinking, I decided to wallow myself in it for a short relief; it was impossible to get up because the heavy heat weighed on our skulls. The screams were frightful, I had to flee but where (did I have to go to)? Whom could I turn to, who could I ask for help? We were all in the same boat, impossible to hold out our hand to anyone... The mud that gave me a short relieve seemed to heathen very quickly, I had to wallow constantly, my lips seemed to open because of the heat. It was a mission impossible to get out of this horror! Nothing guaranteed that this gesture would save my life; it was really by feel that I moved on...Lying down on the ground after having turned on my back to cool it off in the mud, I faced the flames for the first time and I was petrified by the monstrosity that dominated the roof of the factory and I asked myself "What is that thing?" Then, I invoked my grandparents, my parents, my children, my wife; I asked all of them to protect me! I delivered a sermon to God: why had he given me a child in good health three months ago... why was he taking my life at this very moment; it all appeared absurd to me!? I cried "No, no..., I don't want to die! Not this way!" Billions of ideas ran through my mind at that very moment, I saw myself already mummified after the evaporation of all my body water, because we were consumed little by little by the flames; a rather apocalyptic image for the family who would have to identify me..., an atrocity.

The fourth shock: the terrible shock from rescue to despair

Having left the site, I saw on my way a victim propelled by the blast of the explosion. He was trying to drag himself to a safer place; the back of his head was still smoking, his body, scaled like a fish, reflected the torture we were undergoing, so I decided to stop to deliver help... After a few steps, we fell down on the ground, exhausted... At that moment I was forced to save myself, the noise was so frightening, the heat remained unbearable, the perimeter was being consumed... My legs took over the command, automatically, but as by telepathy I could hear that unlucky person asking for help; was it simply my conscience..., I don't know!? So I looked back once more notwithstanding the fear of the pain that hit us..., the decision was unambiguous, I had to help him again; that was the moment he understood that he had to hold out his arms to be pulled into safety. What I lived through was appalling! The skin of the man sticked to my hands, and slipped through my hands... With only little oxygen in the lungs, wearied out, powerless, despaired, I kneeled down to cry! And I said "my God!". I could clearly read the suffering, the fright and the consciousness of this man who was lapsing into a coma without return... unable to complain because he was so in shock! I had to decide quickly but what did I have to do!? I could see the grass smoking around us, I had to get him out of there to limit the damage, the only solution was to put him face down, at least his face had to be saved; I can imagine the triple suffering of this man when executing this manipulation... The poor victim smelt of burnt keratin. Still powerless in the face of such horror I wanted to see with my own eyes the evil that was after us; I stared at the huge fire mushroom, without knowing what it was, or without being able to imagine whether it was the devil or simply the apocalypse according to Saint John!? It was a big shock! A terrible shock! I was still ignorant at this stage about the gas leak; a totally abominable surprise. This moment is for me the quintessence of my traumatic journey; nothing from our birth till our death prepares us to face such an atrocity, unless it is the same kind of event... I had just been embraced, in all her disdain, by the Medusa, without knowing it! Specialists talk about a rise in temperature till 1200°C on site that devilish day!

The fifth shock: the shock of the traumatizing pain

On the way for care, I was immediately put in a car that drove me to a safe and quiet place. When I arrived at that place, my body started to shake due to the fright, the fear of having escaped from a sure death, because those who stayed behind never got out, nearly all of them died, except for one man who stayed petrified inside, hidden under a table. The pain rose gradually and proportionally to the decrease in adrenaline intensity! My nerves were in a state of absolute arousal! I was cold due to the anxiety but the fresh water felt very good... I tried to understand what had happened, now being safe but still not grasp-

ing what was going on. Every passing minute, I could see a change in the survivors of the explosion; the bodies in front of me were transforming, blisters were forming visibly... At a given moment I heard someone say: he is burnt too! The perplexity reached its summit as I still didn't know the extent of the injuries given the horrifying spectacle in front of me... Then I understood my pain! One by one, we were evacuated into a school used as hospital of misfortune, I still remember me banging my head against the ambulance because I was suffering that much...; the mud helmet, as I called my thermal shield, had protected me against the burning of my skull, just like my friends who were also burnt. This state of health kept intact my nerves which made a living skinned man of me! The series continued, this was my fifth shock, also my fifth traumatism!

The sixth shock: the house of suffering and horror

The available rescuers took care of me and I begged them from my arrival on for a tranquillizer because I was afraid I would go mad due to the pain... I received three doses intermittently, very quickly the limit was reached, my heart was at great risk... Moreover I was under monitoring because of an irregular heart rhythm. Around me, there was only crying and suffering..., a real horror museum! The bodies of some victims looked deformed! When you get a strong dose of tranquillizers and you add to that the cries, the smell and the deformed bodies, you lose your sense of reality; you really believe you are taken hostage by Hades, the Greek god of hell. According to mythology, he was the son of Cronos and Rhea, thus the brother of Zeus and Poseidon. After having dethroned their father, Cronos, the three brothers distributed the Universe itself among them! This was the apotheosis of inhumanity that we could live through!

The seventh shock: my mourning after my death, a trauma by correlation

My heart was being monitored, I was still in a daze; the physician at my bedside asked me for the telephone number of my family to reassure them about my condition; bewildered, I gave him the phone number in reversed order; finally he was able to contact my family who was already mourning... My superior had informed my family that, to his regrets, he had no news about my situation on site... My family was devastated by the news! It seemed that my name had escaped the medical staff's notice. The victims who had not been found so far were considered to have passed away! It is easy to understand the relief of my family after the phone call of my treating physician, my benefactor. But it was this same phone call that informed my family that I was severely burnt, traumatized, weakened, but alive! This is why I call it the corollary shock, a shock with a retroactive effect!

The eighth shock: hospitalization and becoming aware

Hospitalized the day after the disaster, badly suffering from the burns, I lived under the influence of tranquillizers for a couple of days, I was spaced out... Four days after my hospitalization, I saw the announcement of the death of a nth victim, not the least; it was the person that I had tried to save... It is only at this moment that a feeling of failure and of culpability invades you. So I underwent an eighth shock! The first tears started to appear... I was not aware that, at that moment, the door to the house "depression" opened for me. I was given a first dose of Xanax to see life through rose-coloured glasses, but quickly I had to switch to Deroxat; a few weeks later, I was dismissed from the university hospital to start a long recovery period together with my family. I had just become father of a son who was three months old at the moment of the drama; the family wanted me to be with them, also to assist my wife who had to be at her work every day...

The ninth shock: the recovery and its "bluff"

I have a vivid recollection of having become the incarnated goodness; avoiding conflicts, showing myself immediately close to others, I had developed a not so rational generosity... During this period, when I listened to some people around me attending evangelical services, I took myself for the resurrected Jesus; my wife invited me to be extremely careful with respect to the intensity of fraternizing with others because she detected a certain anomaly in my habits, not that I repelled others before but now I gave them excessive attention! So I seemed enthusiastic about the idea of living this second life; the unlucky ones died but I did not, so God definitely had to exist and love me, I felt a need for a spiritual search, the thirst of being "impregnated" by the other was there... So I had to get into contact with the professional world again with the simple aim to show up again and also to take stock of the disaster and its drama. All of this was just a deception! I lied to myself; this hyper-availability hid hyper-sensitivity, all it needed to crash was a trigger. My benevolent temperament hid a lot of anxieties; doing people services was a form of seed to harvest a peaceful climate, in reality this motivation hid a simple egoistic interest; actually, it was about recreating "My own Universe" because the other one terrified me and seemed so absurd to me...

The tenth shock: my fall or the dive into the psychiatric abysses

I had an appointment with my superior at Diamant Board (Husqvarna nowadays) – the factory that exploded on the day of the disaster. After a cup of coffee or two, my superior had to announce to me what he had to announce; the parentheses about the professional projects and the family had to come to an end... My superior announced the death of a significant victim, i.e. one of the young bachelor students working under my supervision who hadn't even started to dream about the type of daughter or son he would have wanted to have. The message hit me like an anvil on the head; I knew that the boy was severely burnt, his arm was burnt till the bone, he would lose it according to what was told, but not his life. I had bet on his youth, but that was wrong, he passed away exactly a month after a combat lost against death. I immediately broke down, the sound of my superior's voice reverberated in my ears like unleashed toddlers, I couldn't distinguish anything anymore, everything seemed blurred. The boy had become like a little brother to me after weeks of being severe to him, I would pick him up at Diamond Board Brussels in the morning to drive him to the work site in Ghislenghien. He would talk to me for days about his sick father he wanted to help with his modest contribution to the family. I would express my proudness because he had started academic studies, which was not common in the world of young immigrants, I would encourage him.

The images of the apocalypse scrolled through my mind, my thoughts focused on the Ground Zero; I was thinking about the day before [the disaster] when, after dinner, I went to the roof of the factory to read my newspaper and to enjoy the breathtaking sight on the city of Tournai... and I was thinking of the deflagration associated with my Olympian rest. A cable just snapped in my head! Everything was intermingling in my head, death continued to haunt us! Hades was after us! At this stage of the day, a nuclear fusion of the real world took place, everything was mixed up!

Selection of some concepts useful in psychodynamic theories

From the traumatogenic event to the psychic trauma

In the etymological sense, a "trauma" can be represented in a metaphoric way as a sudden blow to the head; the trauma is then the injury caused by the blow. Interesting to know is that the word "traumatic" stems from the late Greek "*traumatikos*" which means "concerning injuries, good for injuries". This resulted in late Latin in "*traumaticus*", which means "effective against injuries". But *traumaticus* is also derived from the prefix "trauma" which means "wound" in the proper sense of the word, and "damage", "disaster", "deterioration" in the figurative sense.

It appears clearly in S.D.'s testimony that the sudden experience of a colossal deflagration, followed by an extreme heat, lives up to the original description of the trauma concept.

A traumatogenic event is unexpected, violent, unforeseeable and inconceivable. De Clercq (De Clercq & Lebigot, 2001) was the first to use the concept of "*traumatogenic event*" to differentiate between the cause of the psychic trauma, sometimes named "the trauma", and its consequences, also often called "the trauma". He came up with the idea that the psychic trauma is not predictable based on the seriousness of the facts but that it is a real encounter of an individual with the event. In the other testimonies of the Ghislenghien survivors, collected at other moments during meetings – formal as well as informal ones – the consequences of this encounter between the individual and the ultimate danger are found in multiple forms. Notwithstanding this diversity of forms, the traumatogenic event shows stable characteristics. More than being a spillover of the defense mechanisms of the psychic apparatus, the "trauma" is an aggression that a subject undergoes and which is felt like putting in danger its psychological balance. It is a wound in the etymological sense.

The brutality of the confrontation with the realilty of death, present in the terrible Ghislenghien explosion, causes in the survivor a sideration of his ego, leaving him petrified and unable to react for a short while (Crocq, 2002). It is necessary to add to these effects, the omnipresence of the couple of affects dread of death/distress. This dread of death is an extreme fright related to the imminence of death. It is so intense that it causes an experience of great powerlessness, of desperation vis-a-vis the potentially traumatizing situation. On the long run, it can be at the origin of a sustainable psychic change (Lebigot, 2004).

The explosion of the gas column, as a direct vital threat, surprised S.D. in a state of rest. The effect of surprise and incomprehension created a state of powerlessness, confusion and derealization. This traumatogenic event sets, in theory for always, a source of disturbance, an "internal foreign body" in the psychism (Freud, 1951a). The effects of this effraction of the psychism will continue to be felt, even after the threat has disappeared. The psychic apparatus, in the grip of these excesses of stimulation, will do all it can to assimilate these stimuli or, on the contrary, to reject them. These attempts are vain. The event is thus experienced as a foreign body by the psychic apparatus (Freud, 1951a).

According to the Freudian vision on psychological trauma by Barrois (1998), it should be considered that in case of experiences leading to a traumatic neurosis, the external protective stimulus barrier is pierced and that quantities of too large stimulations reach the psychic apparatus, so that here we are facing the second possibility, namely that the fear is not only an affect which announces the danger, but is also produced as a new indication, derived from the economic conditions of the situation.

The repetition syndrome

On the clinical level, after confrontation with a traumatogenic event, signs of stress and fear will initially be felt; later the **repetition syndrome** will appear, witnessing the effraction by the traumatic images. In S.D.'s case, it is about the sensory images of the explosion, odours and images of the burnt or mummified bodies and the consecutive wounds accompanied by extreme pains.

What makes the "trauma" in this hell of fire and heat, is the encounter between an individual and an event beyond the conceivable and the representable; S.D. making an association between the state on the site of the catastrophe and the apocalypse according to St John.

Barrois (1998) proposes a conception of trauma as a break of the link with the world. The individual is confronted with the unthinkable. It is unthinkable to see and to feel oneself pulverized by an explosion, thinking that the heat will lead to a death by evaporation. This confrontation, for which one cannot be prepared, causes the dread and leads to a presentiment of "*the death of oneself as ultimate truth*". The individual is facing the absolute truth of death. This trauma is experienced like the continuity of existence breaking off, the relation to others being suspended. And, even if we are aware of death as our inescapable end, death itself does not have a representation in our unconscious; we do not have the experience of it. Everything happens as if we knew it but without really believing it. This idea penetrates in an intimate confrontation with death, death of oneself or of someone else. There has been a "*putting in danger of the instinct of self-preservation*" composed of the organic and bodily functions allowing the conservation of life (Freud, 1951b).

According to this angle of incidence, the trauma would thus be a way of answering an experience of a break in the meaning of the continuity of one's own existence. The trauma being the link between "*before*" and "*afterwards*", it confronts two radically opposing universes. In this context of disaster, it is about the world of "those taken hostage by Hades" opposed to the world "of the others – the speaking beings".

The "miraculous" survivors of the Ghislenghien disaster did not have any chance to *cope* with their psychism or to see it reverse to the normal because this reversal requires a resolution of the experience lived through by means of neurovegetative symptoms which can die out and of a lucid conscience of having been able to escape from the vital threat of that event. The wounds and burns, driving "insane by pain", testify to this incapacity.

The encounter with the real-of-death

According to Lebigot (2005), the psychic trauma results from an intimate encounter with the *real-of-death*; "real" being understood here in the Lacanien sense of the term. In other words: the psychic trauma is an encounter with the *nothingness*. It is in the nothingness created by this enormous deflagration which seemed to have blown away all the oxygen necessary to survive, that the traumatic image for S.D is represented. However, Freud had already pointed out that death is not represented in the unconscious (Freud, 1951a): we know that we will die, but we do not believe it; we live as if we were immortal. Also, the traumatic images of S.D. cannot find, in his psychic apparatus, the representations to accommodate them. The testimony clearly shows to what level the network of representations is disqualified, unable to transform the real of this explosion into reality. This transformation operation is what we usually carry out starting from our perceptions; it immediately creates meaning and significances (Lebigot, 2005).

Lebigot (2005) describes the incrustration of an image of death in three kinds of circumstances: 1) when the life of the subject himself is threatened and it is certain that he sees his own death (for example, victims trapped in motor vehicle accidents); 2) when the real-of-death is perceived through the death of someone else; this in most cases, but not always, in circumstances where the effect of surprise plays its part; and, 3) when people are implied in the death of someone else and were thus prepared for the death of the other since they are the perpetrators or the active witnesses of it. The three types of circumstances are present in the testimony of S.D. In his experience, the deflagration violently imposed annihilation and this for various reasons: the suddenness, the unexpected, the degree of exposure, the threat of immediate death, the painful journey of physical survival and the identification with the victims.

The dread of death

Thus, the penetration of the traumatic image does not meet any representation in the psychic apparatus (Lebigot, 2005). In the testimonies of the victims, it is precisely *a vacuum of thoughts* and a *complete lack of words* which is accompanied by a *complete lack of emotions* that is experienced. However, S.D.'s experience indicates that the cognitive and/or emotional vacuum is alternated with moments of reflection and decisions to be taken – in a state of disorientation and derealization indeed – but optimizing any chance of survival while facing the fear of death.

Freud had very clearly made a distinction between fear, anxiety and dread. Fear and anxiety can refer to stress. Dread, this moment without affect nor representation and condemned to the image itself, signs the trauma (Lebigot, 2002). In S.D.'s testimony, this distinction between concepts is illustrated by the difference of the experience at the very moment of the detonation itself – i.e the vacuum and the derealization, the world which stands still for a short moment – on the one hand, and the impressions during his "journey of the survivor", under the push of the survival stress accompanied by an extreme anxiety –- i.e anxiety for the monster killing by fire and heat, on the other hand.

Selection of some useful concepts in the empirical theories

The neurobiological premise

Recent neurobiological research describes the neurovegetative storm through which the victims of traumatogenic events go during the intimate encounters with death. The studies of Perry (1999) and of Van der Kolk (1994) show that the brain and the body do not only react to the threat, but are also likely to be modified by the traumatic experience. On the neurobiological level, the integrative failure of an event related to a threat can also appear in the form of an excess of stress hormones and of stress-related deteriorations in the cerebral areas working for the principal integrative functions.

Siegel (1999) explains that each individual manages to preserve the integration of the various sensory stimuli which reach him only between certain limits of arousal. S.D.'s descriptions, evoking intimate encounters with death, illustrate these moments where the level of arousal is too high to remain between the limits of the "window of the tolerance for (neurovegetative) arousal", described by Ogden and Minton (2000). The internal states of the victim, beyond this "**window of tolerance**" (cf Figure 1), are characterized by either an excessive rigidity or a random appearance. These states are inflexible or chaotic, and consequently not adapted to the internal and external environment.

Figure 1. The model of the modulation of neurovegetative arousal (Ogden and Minton, 2000).

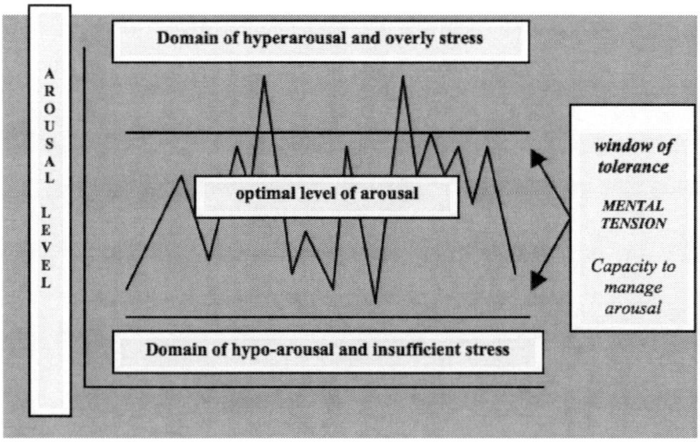

In the states of mind outside this window of tolerance, the *mediating prefrontal capacity* – responsible for the cognitive processes and the handling of information on the level of the brain – allowing to react in a flexible way is temporarily neutralized. The "**higher mode**" of integrative cognitive processes has been replaced by "**lower mode**" reflex processes and sensorimotor processes. On the level of the brain, only the "*animal parts*" of the brain (responsible for the survival behavior) are activated (Siegel, 1999). It is at this stage that the mechanisms of automatic reactions and of peritraumatic dissociation intervene with only one criterion: to survive. On the neurobiological level, the integrative failure of a threat-related event can thus be caused by an excess of stress hormones and by stress-related deteriorations in the cerebral regions working for those principal integrative functions.

Several studies (e.a. McGaugh, 1990) have shown that, during a state of stress, neuro-transmitters discharged in high concentration in regions of the brain that are implied in the execution of integrative mental operations – such as the *hippocampus* and the *prefrontal cortex* – can interfere with the integration of the experiences. These substances, which include noradrenalin, adrenalin, glucocorticoïdes, endogenous opiates and several others, can decrease the level of mental operation of an individual, i.e his integrative capacity. According to these empirical studies, linking the traumatogenic event to the representations will thus be almost impossible due to the produced neurobiological substances.

The various stages of the "traumatization"

During the traumatogenic event – i.e the moment of the deflagration – constituting the "peritraumatic" stage, S.D. as well as the others directly involved showed a significant series of behavioral phenomena as we find them in animals at the moments corresponding to a direct threat to their life. These are behavioral phenomena which will correspond to the various stages of the traumatization by predation. It is possible that the victim of a sudden explosion remains blocked in every one of these stages, and this during the whole evolving traumatogenic event.

The peritraumatic reactions of the people involved at this stage are thus from a phenomenological point of view comparable with the reactions of a prey facing its predator (Nijenhuis, 1999). We find five types of reactions successively: 1) *immobilization* – which in nature sometimes corresponds with the survival chance or with the optimization of the survival chances – and *inhibition*, stiffening or apparent blocking as from the moment when the danger is evaluated (at this stage, one talks about *apprehension of the danger* and eventually about *freezing*); 2) *escape behavior* (at least if there is a way out); 3) *fight when facing the danger* – for example a fire, an accident, an explosion – as far as this "fight" is meaningful for the survival; 4) *complete submission* – the moment when the disaster victims experience an overwhelming powerlessness, broken or unable to defend themselves, and when survival is at stake and they have survived indeed; 5) *restoration* and *recovery*, the stage where the feelings of pain return and one becomes conscious of his wounds. For an innovative and comprehensible outline of this kind of reactions and traumatization, we refer the reader to the work of Nijenhuis, Van der Hart and Steele (Nijenhuis & Van der Hart, 1999; Nijenhuis, Van der Hart & Steele, 2002; Van der Hart, Nijenhuis & Steele, 2006).

Congruence between the psychodynamic and the empirical theories

The theoretical developments above show that these theories of psychoanalytical or psychodynamic inspiration, which are typical for the vision on the psychic traumas in the French theoretical currents, can today be illustrated and/or interpreted by neurobiological research on trauma.

The phenomenology of the psychic trauma as described by Lebigot (2005) corresponds particularly well to the neurobiological phenomena explained in the paragraphs above. According to Lebigot (2005), a subject involved in an exceptionally serious and shocking event is confronted with the question of death; a confrontation which happens according to two modes that, in a private clinic, will not have the same consequences. The first mode is that of the **vital threat**, which will attribute the reactions to "stress" – i.e. the reactions of neurovegetative activation and of adaptation to the threatening external environment which indicate that the individual prepares himself to act against the danger. An adapted and natural reaction then consists in going through the various stages of neurophysiological activation successively; a process by means of which the individual mobilizes himself to face up to the danger and to survive the vital threats which can make him succumb.

The second mode is more complex: the subject has an encounter with the **real-of-death**. According to Lebigot, this encounter is accompanied by (first) **dread (of death)** and (only then) **anxiety** and can, in the aftermath, provoke stress reactions too. It is only later that the appearance of a **repetition syndrome** will testify that there was a psychic traumatization, as illustrated by the experiences S.D. lived through – among others regarding the impressions of his "ten successive shocks" but especially at the time of the deflagration itself. This sometimes forgotten **brief moment of dread**, during or after which the adapted stress increases to an excessive level of stress and activation (*hyperarousal*), puts the victim outside the window of tolerance described by Ogden and Minton (see Figure 1).

In stress situations, the subject faces the threat, mobilizes his defensive resources (from this point of view anxiety is part of it) and succeeds to keep every image of death outside of his psychic apparatus, like he usually does. On the other hand, in the trauma caused by a sudden deflagration, the images of the "real-of-death" have been able to pierce into the psychism and have encrusted there like "internal foreign bodies".

Janet (1889) introduced the term "idées fixes" (fixed ideas) to qualify them and his work is already very clear regarding the concept of psychic trauma, even if it is about so much more than "ideas" only. In these two descriptions, the raw recollection of the event (sensations, suffering, images, immutable thoughts, relived experiences, elementary gestures) – in a metaphorical way called "foreign body" or "fixed idea" – will remain in a remote corner of the subconscious, forgotten by the vigilant conscience and will behave there like a parasite, not assimilated, provoking automatic non-elaborated reviviscences, a source of pathological symptoms. While the remainder of the conscience continues to inspire thoughts and elaborated well-detailed actions, there is "**dissociation** *of the conscience*" (the vigil or alert conscience vs the subconscious).

This traumatic image – "foreign body" (Freud) or "fixed idea" (Janet) – thus penetrates the psychic apparatus and does not find any representation in the unconscious to accommodate it, to bind it, to transform it. Hence, it will not behave like a recollection: it will remain intact, into detail, and when it emerges to the conscience (in a nightmare or when being awake) it is always at present, like an occurring event (Lebigot, 2005).

Janet proposed, also in 1889 already, two different memory systems for the recall of emotionally intense events. According to him, the first memory system is autobiographical; it is verbal and it fulfills a social function, allowing the individual to communicate his experiences to the community around him. The other system would be called today the "implicit memory", which contains the sensory and emotional prints of emotionally intense events (Pitman, Orr, Shalev, Metzger & Mellman, 1999; Van der Kolk & Van der Hart, 1991). According to Janet, these two memory systems function relatively independently of one another.

Today, thanks to neurobiological research, we know that the subcortical areas of the brain – thus the primitive parts which are not under conscious control and which do not have a "linguistic" representation – have other ways of recalling (in particular the somatoform recall) than the more developed regions of the brain, localized in the prefrontal cortex. Under normal conditions, these memory systems are integrated in a harmonious way, but during an excessive stimulation – for example after confrontation with vital threats – the limbic system and the rachidian trunk (brainstem) can produce emotions and sensations which are in contradiction with the normal attitudes and beliefs of an individual, causing "irrational" and "reflexive" behaviour, comparable with the survival behaviour in animals (see above).

Contemporary scientific research (Arnsten, 1998; Birnbaum, Gobeske, Auerbach, Taylor & Arnsten, 1999; Rausch et al., 1996) shows that the high levels of neurovegetative arousal interfere with the functions of the frontal lobe (a.o. the area of Broca) necessary to verbalize the experience of the trauma. During the repetition syndrome, when the original traumatogenic event is relived, the brain functions differently than at moments of avoidance; precisely this makes the integration of the event extremely difficult and turns traumatic recollections into a set of fragmented sensory and emotional prints (Van der Kolk & Fisler, 1995; Van der Kolk, McFarlane & Weisaeth, 1996a; Van der Kolk, Pelcowitz, Roth, Mandel, McFarlane & Hermann, 1996b).

Discussion: the psychic trauma of the survivor of a catastrophe (by explosion)

An integrative theoretical explanation?

The neurobiological explanations given above correspond particularly well to the psychoanalytical and hermeneutic trauma theories according to which the psychic apparatus remains parasitized by the effraction. The traumatic anchoring can be an image or a noise: mummified bodies of dead victims and odors accompanying these images. Then the psychic trauma shows a clinical set of pathognomonic symptoms and of nonspecific symptoms. Its pathognomonic sign is **the repetition syndrome**. This syndrome marks many attempts of adaptation to the peri-traumatic moment. This repetitive and traumatic dimension is an attempt of the traumatized person to find a connection, a symbolization of the event (Crocq, 2000).

In 1889, Janet stipulated that the "vehement emotions" at the moment of the traumatization make it practically impossible to find a satisfactory narrative form for the experience of the traumatized survivor. Thus, Janet was the first to say that the incapacity of the victim to perceive and to *personify* his whole experience would cause a failure on the level of the memory: instead of being organized like a set of coherent recollections, the event becomes a complex of disconnected emotions, of visual perceptions and/or kinesthetic sensations, which can re-appear on the surface when the victims are confronted again with sensations and/or emotions that recall the original event (Van der Hart, Nijenhuis & Steele, 2006). Janet (1889) defined mental health in terms of integration capacity. He contended that integration requires the continuous execution of a series of mental acts. With a reduced integrative capacity, the individual undergoes an immediate deficit of his capacity to adapt to the trauma, and the development of his capacities to face later traumata is inhibited.

The memory of the traumatized subject is thus caught in a trap by the event, like the memory of S.D. who was taken hostage by the many triggers that were omnipresent due to the media attention regarding the Ghislenghien disaster and by the suffering of the severely burnt person in the long run. Each medical treatment provokes reviviscences in daytime as well as at night. While sleeping, repetitive nightmares occur in which victims are begging for assistance, an analgesic against the pain, a tranquillizer against the suffering. These are always different editions of the same basic scenario, the latter being the event that has occurred, i.e. the demolishing deflagration. This repetition is source of anxiety but also source of fascination (Freud, 1951).

The experience of something exceptional like the trauma makes that the traumatized person is secretly fascinated by this initiation. That phenomenon appeared very clearly in the disaster victims of the technological catastrophe in Ghislenghien. During this catastrophe many photographs were taken. Quickly, these photographs were duplicated so many times that each family of a victim could have their own photograph and/or film gallery.

There have been many meetings between people involved in the catastrophe. The history of the catastrophe starts to belong to the collective memory of the victims involved but also of their families. Pride, solidarity, friendship, admiration, desire and fear are the words that described the atmosphere during those meetings of victims, but also anger and indignation regarding the responsible authorities who have abandoned the people involved in the Ghislenghien disaster for a long time.

The feeling to have become different

The experience of confrontation with the real-of-death gives S.D. the feeling that he has become different. Initially, in his euphoria of miraculous survivor – as he explains in the description of his ninth shock – he believes he's in the dream, the delirium or rather the illusion of new life with a spiritual destiny. But while returning to the same level of functioning as his close relatives, after a few months and still nowadays, he has the feeling that he is *completely transformed*. He often feels more prone to fatigability and irritability. His level of attention and concentration is significantly lower and his intellectual activity burdened with memory lapses because the Ghislenghien disaster is always omnipresent, requiring an impressive quantity of energy every day.

Other Ghislenghien victims also said to have "the head in the clouds" and sometimes worse: they had to think deeply before being able to give their date of birth or those of their children to the insurer. These memory lapses were experienced as very painful. Let us not forget that a neurobiological reality is hiding under these psychological phenomena: the reality of an organism which has been in a state of alarm, uptight and hypervigilant, and which has mobilized itself to resist until exhaustion or total collapse to the sometimes vital threat. A brain that, during the confrontation with the real-of-death, was overactive and this sometimes unilaterally; i.e. the right lobe was recording the preverbal sensations and the raw images of the traumatogenic event, while the left lobe (responsible for the handling of information, the vocabulary and the verbalization of emotions) was practically inactive.

The sensations, raw images and basal emotions remain thus "non-metaphorized" and "non-symbolized".

The feeling of abandonment

During the meetings the other Ghislenghien victims told that they considered the initial feeling of abandonment as central in their experience. The traumatic experience - the deflagration – was experienced as incommunicable, inexpressible. The dread, with the potential to cause the hyperarousal of the neurovegetative system, and thereby to reduce the experience to a set of raw images and sensations (somatoform and psychoform), may "disconnect" the left lobe of the brain. This makes it impossible for the traumatic image to find a representation in the psychic apparatus, remaining cut off of any meaning. To thwart this feeling of being abandoned, traumatized people may show an important reduction of their social contacts aiming at a progressive containment at home.

The feeling of guilt

The culpability of the survivor appears clearly in S.D.'s testimony: the fact of surviving without having been able to help other victims. Survivors of catastrophes continue to wonder: *"Why did we come alive out of it while so many others died?"* or *"Why didn't we have more opportunities to help other victims who needed it?"*. Sometimes this culpability is projected on people supposedly guilty of the consequences of the catastrophe. Hence, it is sometimes more *comfortable* to seek responsibility, incarnated by humans, rather than to accept that in fact the accident or the catastrophe was caused by coincidence. This culpability is related to the metaphor used by Crocq (2000) "having returned from hell", and to the idea of having been, in spite of oneself, in contact with a prohibited primal knowledge (Lebigot, 2005).

In conclusion, and after Lebigot (2005), it is important to stipulate that *"together with the affect of dread, it is the experience of absolute loneliness, of breaking off all the community and cultural bonds that characterizes the psychic trauma."* Everything occurs as if the trauma induced a process of precariousness.

The development of problems around culpability and shame is also often one of the peculiarities of the person who experienced a trauma, and who tends to crystallize it. These problems hamper the request for care. In fact, the dominating feeling is one of inability to be understood in what was experienced, or sometimes even the incapacity of being listened to. Facing that much incomprehension – real or perceived – the disaster victim who suffers from a trauma *"gives up his attempts for recognition by the interpreted and interpretable world"* (Douville, 2001). In this respect, it is interesting to copy the words of another Ghislenghien survivor, who expressed his feeling of total isolation in these terms: *"Even when having the nicest person in front of me to talk to, when I speak, it feels like speaking to a Russian. We do not speak the same language anymore."*

Conclusions and recommendations

In this article, the rendering of S.D.'s experiences, survivor of the terrible explosion in Ghislenghien, was analyzed in the light of various theories about the psychic trauma; on the one hand, theoretical and clinical psychodynamic descriptions, and, on the other hand, current neurobiological concepts.

This work makes it possible to apprehend the hypothetical possibilities of "field support" (during the peritraumatic stage) in case of collective urgencies or catastrophes and this by listening carefully to the first stories of victims. Before being able to deliver this field support, it is paramount to thoroughly understand and to master "the anatomy" of the traumatogenic event and its consequences.

How to understand the constitutive factors of the concept of dread and how to measure the adequacy of peritraumatic support interventions in the field? What would be the measuring criteria of this efficiency? Indeed, in the work of first rescue workers and psychosocial caregivers in times of disaster, it is important to be able to understand the experience lived through by the miraculous survivors in order to be able to provide an adequate first immediate and acute support.

The objective of the work is thus centered on this single encounter with the victim in the peritraumatic stage. The metaphor that can explain the work of the first contact is that of a *lifebelt*. The main aim of that *lifebelt* is to ground the detained victim in the reality, during the rescue operations, to thwart the feeling of total abandonment, "to fill" the nothingness the traumatized victim is coping with, to keep this victim in the "window of adaptive arousal" (see Figure 1), and to minimize, as far as possible, all vehement negative emotions, pains and forms of peritraumatic dissociation. Amongst other things, the question is to seize the unique opportunity to quickly deliver field support to the traumatized victims or to those in the traumatization process; and this as from the initial moment of the accident until the moment the victims are brought in safety or arrive at the emergency department. By doing so, peritraumatic dissociation and sometimes persistent dissociation consecutive to the confrontation with the real-of-death is prevented (Panasetis & Bryant, 2003). It is at this moment of horror and petrification, that the victims are at risk of experiencing inexpressible impacts, called in this text, "right lobe experiences", causing constellations of sensations and raw images, which are impossible to grasp in words. When this field support is delivered, it is also necessary to accompany the victims in their reactions of psychic and/or physical survival.

The problems induced by collective emergencies or disasters, which cause a great suffering to the people who are exposed to a threatening confrontation and who are often forced to remain immobilized in a feeling of total abandonment, being confronted with the real-of-death, justify the attempts of first psychological help aiming at making it possible to verbalize the experience. It was made clear above that the verbalization of the experience itself can already be extremely difficult. Moreover, when the latency period is fading out and gives way to the emergence of traumatic symptoms, there will be no more means to erase the traumatogenic impacts which have taken place during the first moments of the experience.

Authors' contributions

EDS conceptualized this article and interpreted the experiences of the disaster survivors according to French trauma theory and current neurobiological theories. He drafted the manuscript of this article and collected relevant literature references. JM critically reviewed the document for intellectual content and style before giving his final approval for this version to be published.

Acknowledgements

The authors would like to thank S.D. for his militant actions in profit of the collectivity of Ghislenghien victims, apparently forgotten by the Belgian authorities as well as by the many "trauma specialists" who showed up in the immediate aftermath of the disaster. The courage and the personal investment of S.D., often to his own detriment, have made the elaboration of this testimony possible and gave so many others involved a voice again. The authors would also like to thank the Professors Onno van der Hart and Rolf Kleber of the Faculty of Social Sciences at the University of Utrecht (the Netherlands) for their availability and their recommendations on the drafting of the articles concerning the Ghislenghien disaster. Birgit Mylle and Aurélie De Blomme have been very helpful in the translation of this article.

References

Arnsten, A.F. (1998, June 12). The Biology of Being Frazzled. *Science, 280,* 1711-1712.

Barrois, C. (1998). *Les névroses traumatiques.* 2nd edition. Paris: Dunod.

Birnbaum, S., Gobeske, K.T, Auerbach, J., Taylor, J.R., &, Arnsten, A.F. (1999). A role for norepinephrine in stress-induced cognitive deficits: alpha-1-adrenoceptor mediation in the prefrontal cortex. *Biological Psychiatry,46,*1266-1274.

Crocq, L. (2000). Le retour des enfers et son message. *Revue francophone du stress et du trauma, 1(1),*5-19.

Crocq, L. (2002). *Traumatisme psychique et névrose traumatique.* Cours du Diplôme Universitaire Stress & traumatismes majeurs. Paris.

De Clercq, M., &, Lebigot, F. (2001). *Les traumatismes psychiques.* Paris: Masson.

De Soir, E., Zech, E., Versporten, A., Van Oyen, H., Kleber, R., Van der Hart, O., & Mylle, J. (2008). Prédiction de l'ESPT lors d'une catastrophe technologique. *Revue francophone du stress et du trauma,8(3),* 211-224.

Douville, O. (2001). Incidences subjectives des situations de grande exclusion. *Le journal des psychologues,191,*53-57.

Freud, S. (1951a). Considérations actuelles sur la guerre et sur la mort. In: *Essais de psychanalyse.* Paris: Petite Bibliothèque Payot, 1981,7-39.

Freud, S. (1951b). *Inhibitions, symptômes et angoisse.* Paris: Presses Universitaires de France, 1986.

Graux, C., & De Soir, E. (2005). Evénements traumatogènes et psychologues « hors murs » ... à pratique « hors normes ». *Revue francophone du stress et du trauma,5(2)*,109-115.

Janet, P. (1889). *L'automatisme psychologique.* Paris: Alcan.

Lebigot, F. (2002). L'effroi du trauma psychique. Le regarder en face ou s'en protéger. *Revue francophone du stress et du trauma,2(3)*,139-146.

Lebigot, F. (2004). Le traumatisme psychique. *Revue francophone du stress et du trauma,4 (1)*,5-11.

Lebigot, F. (2005). *Traiter les traumatismes psychiques. Clinique et prise en charge.* Paris: Dunod.

McGaugh, J.L. (1990). Significance and remembrance: The role of neuromodulatory systems. *Psychological Science, 1*,15-25.

Nijenhuis, E.R.S. (1999). *Somatoform dissociation: phenomena, measurement, and theoretical issues.* Assen: Van Gorcum, Reprint Norton, New York, 2004.

Nijenhuis, E.R.S., &, Van der Hart, O. (1999). Forgetting and reexperiencing trauma: from anesthesia to pain. In: Goodwin J.M., Attias R. (Eds). *Splintered reflections: images of the body in trauma.* New York: Basic Books, 1999,39-65.

Nijenhuis, E.R.S., Van der Hart, O., &, Steele, K. (2002). The emerging psychobiology of trauma-related dissociation and dissociative disorders. In: H. D'haenen, J. A. den Boer, H. Westenberg & P. Wilner (Eds.), *Textbook of biological psychiatry* (pp. 1079-1098). New York: Wiley.

Ogden, P., Minton, K. (2000) Sensorimotor psychotherapy: One method for processing trauma, *Traumatology*,6.

Panasetis, P., & Bryant, R. (2003). Peritraumatic versus Persistent Dissociation in Acute Stress Disorder. *Journal of Traumatic Stress, 16*,563-566.

Perry, B.D. (1999). The memory of states: How the brain stores and retrieves traumatic experience. In: J. Goodwin & R. Attias (Eds.), *Splintered reflection: Images of the body in treatment* (pp. 9-38). New York: Basic Books.

Pitman, R.K., Orr, S.P., Shalev, A.Y., Metzger, L.J., & Mellman, T.A. (1999). Psychophysiological alterations in post-traumatic stress disorder. *Seminars in Clinical Neuropsychiatry, 4*,234-241.

Rauch, S.L., Van der Kolk, B.A., Fisler, R.E., Alpert, N.M., Orr, S.P., Savage, C.R., Fishman, A.J., Jenike, M.A., & Pitman, R.K. (1996). A symptom provocation study of posttraumatic stress disorder using positron emission tomography and script-driven imagery. *Archives of General Psychiatry,53(5)*,380-387.

Siegel, D.J. (1999; p254-255). *The developing mind: toward a neurobiology of interpersonal experience.* New York: Guilford.

Van der Hart, O., Nijenhuis, E.R.S., &, Steele, K. (2006). *The Haunted Self: Structural Dissociation and the Treatment of Chronic Traumatization.* New York: W. W. Norton & Company.

Van der Kolk, B.A., Van der Hart, O. (1991). The intrusive past: The flexibility of memory and the engraving of trauma. *American Imago, 48*,425-454.

Van der Kolk, B.A. (1994). The body keeps the score: memory and the evolving psychobiology of posttraumatic stress. *Harvard Review of Psychiatry, 1*,253-265.

Van der Kolk, B.A., & Fisler, R.E. (1995). Dissociation and the fragmentary nature of traumatic memories: overview and exploratory study. *Journal of Traumatic Stress, 8(4)*,505-525.

Van der Kolk, B.A., McFarlane, A.C., & Weisaeth, L. (Eds.). (1996a). *Traumatic stress: The effects of overwhelming experience on mind, body and society.* New York: The Guilford Press.

Van der Kolk, B.A, Pelcowitz, D., Roth, S, Mandel, F, McFarlane, A.C., & Herman, J.L. (1996b). Dissociation, somatization and affect dysregulation: The complexity of adaptation to trauma. *American Journal of Psychiatry, 153,* 83-93.

CHAPTER 3

The Concept of Psychological Trauma in Contemporary French Theory: An Unknown Story?

E. De Soir, R. Kleber & J. Mylle

Submitted

The Concept of Psychological Trauma in Contemporary French Theory: An Unknown Story?

Erik De Soir, Rolf Kleber, & Jacques Mylle

Summary

This chapter examines the conceptualization of psychological trauma in contemporary French theory and the clinical importance this approach still holds beyond the community of English-speaking trauma clinicians. Starting with the historical background of the current psychodynamic French concept of trauma, it investigates the potential contribution from contemporary French authors to current knowledge on psychotraumatology internationally and aims to clarify differences and similarities between these approaches in the light of the leading paradigms of today. The current position of the most influential contemporary French authors is that stress and trauma do not necessarily fit into the same concept. The French clinical description of psychological trauma is based on the concept of *effroi de la mort* (dread of death) and the *syndrome de répétition* (repetition syndrome). A better understanding of these clinically relevant but largely unknown theories sheds another light on the currently used concepts in mainstream literature and the discussion on PTSD in DSM-5. The comparison of trauma concepts in different cultures and/or language groups, in a context of globalization of trauma societies, could be useful bridging existing conceptual gaps, understanding critical differences in clinical practice and offering a more integrative view of psychological trauma.

Keywords: PTSD, DSM-5, traumatic neurosis, repetition syndrome, effroi de la mort

Introduction

Scientific research in recent decades has provided a clear, evidence-based description of the serious consequences of trauma-related experiences (Breslau & Davis, 1992; Carlson, Furby, Armstrong, & Shlaes, 1997; Foa & Rothbaum, 1998; Horowitz, 1986). Various concepts have been developed to explicate the consequences of these experiences (Gersons & Carlier, 1992; Kinzie & Goetz, 1996; Wilson, 1989, 1994). *Post-Traumatic Stress Disorder* (PTSD) and other mental disorders (including anxiety, depression, substance abuse, dissociative disorders and psychosomatic disorders) can result from psychological trauma (Brewin, 2000). *Acute Stress Disorder* (ASD) and *Post-Traumatic Stress Disorder* (PTSD) are described in DSM-IV-TR (APA, 2000) as mental disorders which may occur after traumatic experiences such as severe accidents, rape, torture, violence and war. The risk for PTSD increases with the degree of exposure to trauma and both the frequency and the severity of the traumatic events (e.g. Bramsen, 1995).

Over the years, a scientific consensus has appeared in the Anglo-American scientific literature about the symptoms of both acute and posttraumatic stress – i.e. intrusive recollections of the traumatic event, a profound sense of numbness, avoidance, increased arousal and hyperactivity and/or exaggerated startle response and dissociative reactions (Foa, Riggs & Gershuny, 1995). If these reactions – intrusive reexperience, avoidance and hyperarousal – persist (at least for one month) causing significant distress or loss of function, PTSD can be diagnosed according to the DSM-IV-TR (APA, 2000).

In DSM-5 PTSD was included in a new section on trauma- and stressor-related disorders. This move from DSM-IV-TR, which addressed PTSD as an anxiety disorder, is among several changes approved for this condition that is increasingly at the centre of public as well as professional attention (APA, 2013). In PTSD (APA, 2013), the trigger is exposure to actual or threatened death, serious injury or sexual violation. The exposure must result from one or more of the following scenarios, in which the individual: directly experiences the traumatic event, witnesses the traumatic event in person, learns that the traumatic event occurred to a close family member or close friend (with the actual or threatened death being either violent or accidental); or experiences first-hand repeated or extreme exposure to aversive details of the traumatic event (not through media, pictures or movies unless work-related). The disturbance, regardless of its trigger, is supposed to cause clinically significant distress or impairment in the individual's social interactions, capacity to work or other important areas of functioning and was not the result of another medical condition, medication, drugs or alcohol.

Significantly different perspectives, largely dependent on historical and cultural differences, exist in the international trauma literature. Among these, the contemporary French conceptualization remains largely unknown to the Anglo-Saxon field of traumatic stress studies. While this mainstream literature often serves as the reference point for young non-English speaking researchers, the viewpoints of contemporary French-language authors continue to provide the basis of clinical practice for trauma clinicians in France, Belgium, Italy, and Switzerland. Moreover, French concepts also remain highly influent in Latin countries across Europe and South-America. Clinicians and researchers continue to be inspired by classical hermeneutical interpretations, psychodynamic theories and clinical experience. The depth and breadth of the theoretical and conceptual gap which exists between trauma concepts in different language groups is striking. While the existing professional societies for traumatic stress studies struggle to create greater openness to cultural diversity, to our knowledge no attempts have been undertaken to define or integrate these very different descriptive and conceptual views or investigate their relative usefulness for clinical practice.

We consider it important to present current French trauma theories along with the conceptual trauma terminology used by influential contemporary French authors along with a brief review of the historical background from which these ideas arose. The international trauma community is currently adapting to the revised status of PTSD in DSM-5 and a global initiative meant to bridge key differences among trauma experts worldwide has been initiated by the *International Society for Traumatic Stress Studies* (ISTSS). Within the ISTSS, a Global Initiative Project team aims to broaden the international trauma community and to create more openness for cultural diversity. Therefore, the objective of this article is to explore how contemporary French trauma theory might contribute to a more integrative view of the variety of terminologies and concepts used by trauma practitioners in different language groups.

Historical Background

From Ancient History to the Traumatic Neurosis

The history of French trauma theory goes back to the concept of traumatic neurosis as originally defined by the German psychiatrist Oppenheim (1884). Oppenheim was one of the first to identify this post-traumatic disorder as an independent entity in a study on the psychological sequelae caused by fear experienced during train accidents. In many accident victims who had suffered no injuries, or only minor ones, symptoms of psychic deterioration appeared after some time and these frequently resulted in total disability. John Eric Erichsen (1875) published a series of lectures on railway spine which was successful in linking the term to a physical theory. According to Erichsen (1875), the violent jolt of an accident resulted in concussion of the spine. The theory which Oppenheim formulated posited that post-traumatic nervous symptoms constituted a distinct diagnostic entity and resulted from the direct – anatomical or psychic – effects of traumatic experiences. His suggestion was opposed by Charcot (1887) who saw 'traumatic neurosis' as nothing more than a particular etiologic form of hysteria, or – at the most – of 'hysteroneurasthenia' and prompted his students to write a number of theses defending his views. Charcot (1825-1893) studied trauma during the second half of his career, from the later 1870s through to his death in the early 1890s. By this time, he had already completed his major clinical and scientific work in neurology and was a figure of international fame throughout the world of Western medicine. Included among Charcot's voluminous clinical publications are approximately twenty detailed case studies that carry the primary diagnoses "névrose traumatique", "hystérie traumatique", "hystero-traumatisme", or "hystero-neurasthénie traumatique" (Micale & Lerner, 2011). Clinically, what Charcot observed in his practice during the 1880s were curious syndromes following a diversity of (sometimes minor) bodily injuries, marked by disabling physical and psychological features but in the complete absence of any indication of structural damage. The most common symptoms seemed to be motoric and sensory disturbances – anesthesias, hyperestesias, paralyses, and contractures of all kind. Fatigue, headache, back pain, heart palpitations, chest pain, irregular pulse rate, constipation, dizziness and fainting spells, and the trembling of the hands and legs also occurred frequently. Emotional troubles could be part of the symptom profile as well: depressive states, sleep disorders (including insomnia and nightmares), phobias, mental confusion, and lowered intellectual efficiency. At times, these symptoms disappeared suddenly and spontaneously, in a matter of hours. At other times, they persisted for months or even years (Micale & Lerner, 2011).

From Graeco-Roman times onward, conversion hysteria has been well documented. A well-known example is the case of the Athenian warrior who fought the Marathon battle – mentioned by Herodote – and became 'blind' for the rest of his life because he had been frightened by the sudden sight of a colossal enemy whom he had taken for a spectre (Crocq, 1999). Another famous case, mentioned by Pinel (1798), is that of the philosopher Pascal, who – after being nearly thrown with his coach into the Seine river by his runaway horses – suffered from repetition nightmares and could fall asleep only after placing a chair near his bed to reassure himself that he was not on the brink of void.

The cases of 'nostalgia' in campaigning armies during the 18th century and the syndromes of 'cannonball wind' exhibited by Napoleon's armies probably included a large number of (post)traumatic stress reactions, named 'traumatic neurosis' in classical literature. Pierre Janet (1889), and then Freud (1893), looked into the pathogenic role of psychological trauma and memories 'forgotten by consciousness' of traumatic events and 'moral emotions', and each contributed to the discovery of the unconscious and of the 'cathartic' method for curing these disorders. For Freud, Charcot's cases opened the way toward the possibility of a purely psychological explanation of physical symptoms, that is, a theory of conversion, and, beyond this, to a general psychological theory of the neuroses based on Charcot's clinical writings on hysteron-traumatic paralyses (Freud & Breuer, 1893).

Etiologically, Charcot believed that his cases resulted from the combined action of a hereditary *diasthèse*, or constitutional predilection to nervous degeneration, and an environmental provoking agent *(agent provocateur)*. Throughout the second half of the nineteenth century, a doctrine of hereditary determination dominated French mental medicine (Dowbiggin, 1985). According to this view, which Charcot endorsed unreservedly, nervous and neurological diseases manifested a latent flaw or defect of the nervous system – a *tare nerveuse* – that at all times was waiting to be activated by appropriate circumstances. In Charcot's medical thinking, traumatic stimuli acted on this prior constitutional susceptibility, and the fact that some individuals developed elaborate neurotic symptoms following a trauma while others did not was explained by the presence or absence of this background (Bertrand, 1990).

According to Freud's initial psychoanalytic theory, hysteria has traumatic origins (Freud, 1956). Hysteria is seen as a neurotic state in which an experience cannot be admitted into consciousness and is instead expressed in a body part. This is what Freud (1956) referred to as the *deferred effect* or *Nachträglichkeit*; that is an event is not experienced as traumatic at the time at which it occurs but becomes so later once it is given a meaning within the frame of mental representations. This experience cannot be admitted into consciousness and is only expressed in a symptom. In his two most important early works, *L'Automatisme psychologique* (1889) and *l'Etat mental des hystériques* (1893), Pierre Janet (one of Charcot's most influencial students) focuses on the mental, rather than the neurological, stigmata of the neuroses, including phobias, abulias, obsessions, and states of "dual" or "double consciousness". Many of these symptoms result from emotional traumata the memory of which became unconsciously fixed in the patient's mind causing a weakening of the ability for mental and emotional synthesis. The vocabulary (*obnubilation of consciousness, dissociation of the ego*) used by Charcot in the case of "Le Log" – a 29-year-old Breton who had been crossing the *Pont des Invalides* in Paris and sideswiped with a wheelbarrow by a passing horse-drawn carriage – is close to the language of early Janetian psychology.

War neurosis

Actually, the two World Wars were the most instrumental events in establishing the existence of traumatic neurosis, in its etiologic form of 'war neurosis' (*névrose de guerre*), underlining the role of a 'shock-emotion' and emphasizing the 'post-emotional' as opposed to the 'post-commotional' etiology or the 'war psychoneurosis' (*psychonévrose de guerre*) as described by Roussy and Lhermitte (1919). The concept of the 'commotion', while already sounding psychological, is intended in a pure physical sense: a nineteenth French medical dissertation (Delcasse, 1834) defined it as 'the shock experienced by certain parts of the body on the occasion of falls or when being stricken'. The invariable cause of such a 'commotion', the text went on, was 'a shock that affects the same part or a contiguous part of the body to be affected'. The phenomenon of accident shock, i.e., a traumatization of the victim without discernible physical injury, naturally existed before the English medical profession started devoting its systematic attention to it in the mid-1860s, after the passage of the liability laws which in an extended sense also included emotional shock reactions to bad news, experienced catastrophes, etc. French medical writing about *la neurologie de guerre* (*war neurology*) during World War I was largely inspired by Charcot's ideas on the neurology of human trauma. Suddenly confronted with outbreaks of psychogenic paralysis, blindness, and amnesia among soldiers at the front lines, French physicians returned to the ideas of Charcot. Publications in military psychiatry medicine in France between 1914 and 1918 repeated the fin-de-siècle debate about the origins of the traumatic neuroses. Neurophysicians during World War I increasingly came to acknowledge that so-called shell-shocked soldiers suffered not from the direct concussive effects of exploding shells and poisonous gas but from extreme levels of fear, anxiety and fatigue. The war seemed to illustrate on a huge scale Charcot's emphasis on the terribly pathogenic potency of fear (Cygielstrejch, 1916). Many wartime doctors openly characterized shellshock as *hystérie de guerre* (Lefebvre & Barbes, 1984). The 'infantile' or 'regressive' attitude of the war neurosis patients was also mentioned (Charpentier, 1917; Ferenczi, Abraham, Simmel & Jones, 1921), and Freud (1918) as well as the American authors of World War II (Kardiner & Spiegel, 1947) revealed the 'conflictual' interweavings (war superego against peacetime superego, fight or flight) entwined with the 'fright' and 'surprise' dimensions of the trauma.

Conflictual etiologies were again identified in subsequent studies of post-1945 wars and guerrillas, including the French campaigns in Indochina and Algeria (Crocq, 1999), the Vietnam War (Jones, 1982) and the Middle-East conflicts (Belenky, 1987). Renewed interest in traumatic neurosis and related disorders was generated by the development of victim psychopathology – in victims of the Holocaust, rape, accidents or abduction and disaster psychiatry (immediate and delayed sequelae of accidents and disasters) (Wilson et al., 1988).

This historical background makes clear that the original French idea of trauma differs from the psychiatric concept as described in the Diagnostic and Statistical Manual of Mental Disorders (DSM) (APA, 1980; 1984; 1994, 2000, 2013). According to the DSM, an overwhelming event can only be traumatizing when it is observable by others and directly and causally linked to the disorder. In contrast to the latter, the psychoanalytic idea proclaims that events not directly observable by others may also be traumatic.

Psychological trauma in contemporary French literature: the work of Louis Crocq

Louis Crocq and his perspective on traumatic neurosis

Louis Crocq, who was the recipient of the *2006 Lifetime Achievement Award of the International Society for Traumatic Stress Studies*, should be considered as the most influential contemporary French trauma theoretician, closely followed by his contemporary colleagues François Lebigot and Claude Barrois. Crocq (1999) tried to fulfil a bridging role between the classical French trauma interpretation and the current view on psychological trauma in mainstream literature. He pointed out that posttraumatic stress disorder (PTSD, APA, 1994, 2000), except in some etiological differences, corresponds closely to the former 'traumatic neurosis' in the classification of neuroses. Freud, who devoted memorable pages to this condition as early as 1893-'95 (*Studies on Hysteria*), then shortly after World War I, in 1921 (*Beyond the Pleasure Principle*), never disclaimed its existence since he recognized – towards the end of his life – in 1938, that traumatic neuroses had always eluded the infantile conflict hypothesis. In his writings, he elevated the medical idea of trauma from secondary to primary etiological status and linked trauma to the notions of psychosexual motivation and unconscious repression (Micale, 2001).

Traumatic neurosis is recognized by Crocq (1999) as an independent neurosis, i.e. as a structured and lasting neurotic disorder, including both specific symptoms (symptoms of repetition) and general neurotic symptoms such as anxiety and asthenia, and involving a typical organization (which in this case is a reorganization) of the personality which produces and perpetuates symptoms.

Crocq (1999) presents the traumatic neurosis as a persistent residual state that should not be mistaken for an immediate stress reaction. It develops after a latent period that varies across situations and individuals (from several days to several months). This latent period, previously called 'incubation', 'meditation', 'rumination' or 'contemplation' by trauma clinicians, corresponds to a dynamic readjustment of personality, which must elaborate new defences in order to cope with the intrusion of an event that the subject was unable to control at the time of its occurrence. During this latency period, which is apparently silent, the knowledgeable observer can nevertheless detect various signs or symptoms including self-isolation, solitary mental rumination, depressive withdrawal, or, on the contrary, paradoxical and voluble euphoria predicting subsequent problems.

The repetition syndrome in traumatic neurosis

Every neurosis is characterized by its constellation of symptoms: anxiety attacks in anxiety neurosis, specific fears in phobic neurosis, conversion and/or dissociation in hysteria, rites and obsession in obsessive-compulsive neurosis. In traumatic neurosis, the characteristic finding is a set of symptoms related to what is called the 'repetition syndrome'.

Crocq (1999) describes the repetition nightmare as the most important feature in the disturbances after a traumatic experience. The repetition nightmare is experienced rather than contemplated; the traumatized subject relives – under the original form or as a transposition developed to the dream-formation rules of displacement, condensation, and symbolization – the inaugural traumatic scene that overwhelmed his defences. The repetition nightmare is experienced intensely; the subject screams, struggles, occasionally falls out of bed and wakes up, terrified and covered with sweat.

The repetition syndrome also includes other forms of repetition, including resurgence of irresistible thoughts recalling the traumatic scene, mental ruminations, fleeting, near-hallucinatory visions, an irresistible tendency to look for and contemplate scenes of violence in real life or in visual reproductions, impulsive motor behaviours improvising motions of defence or aggression, and, last but not least, startle reactions – either in response to minimal stimuli or spontaneously – that may be considered as the most archaic form of repetition.

The traumato-neurotic personality

Every neurosis is defined not only by its symptoms but also by the pattern of the underlying personality: an anxious personality in anxiety neurosis, and, phobic, histrionic and anal personalities in phobic, hysteric, and obsessive neuroses. Crocq (1999) argues that, in the traumatic neurosis, the underlying personality pattern is not 'constitutional' or acquired during childhood, but rather what the personality has become after the impact of trauma; fearful, inhibited, regressive, dependent, but also contentious and demanding attention and rehabilitation.

Psychological trauma in contemporary French literature: questioning terminology

Contemporary French authors (Barrois, 1998; Crocq, 1992, 1999, 2001; Lebigot, 2001, 2002, 2005) have questioned the terminology of '(post)traumatic stress' used by anglophone authors, claiming that stress and trauma are two distinct concepts and that the most immediate human reaction to a traumatizing stimulus is a typical state of petrified shock or dread. In French literature (ibid) this state is referred to as *effroi*, close to Kraepelin's old concept of *Shreck* (1899). According to this view, intense fear and stress reactions only follow after this first and basic reaction of shivering tremor. In his clinical cases, Charcot conjectured that such a state induced a kind of "somnambulism" or "hypnotic trance" in which there is an increased psychological suggestibility which leads to an reproduction or imprinting of the traumatic sensation as a mental representation. A "traumatic idea" or *idée fixe* is then lodged in the mind of the individual. It is interesting to note that, according to the Oxford English Dictionary, the word "trance" derives from a Middle French word for "being frightened to death".

In line with this view, these French authors have proposed a clinical uncoupling of psychotraumatic syndromes – i.e. syndromes directly linked to the potentially traumatizing impact of a sudden, massive and/or brutal confrontation with death – and the psycho-emotional consequences of disturbing events and life crises (Crocq, 1992, 2001). The accent is put on the unique personal truth of the traumatic experience, which can never be reduced to merely experiencing an overwhelming event and a consecutive biophysical reality of extreme stress (Crocq, 2001; De Clercq & Lebigot, 2001; Lebigot, 2001). In this view, the immediate, post-immediate and delayed stages of the reaction to a potentially traumatizing event are clearly distinguished and therefore a disorder (previously described as 'traumatic neurosis'), for which the repetition syndrome (*syndrome de répétition*) is the pathognomonic expression, can be clearly identified. This interpretation is strongly inspired by psychoanalytic theories based on concepts such as the traumatic unconscious (*inconscient traumatique*), psychological automatism (*automatisme psychologique*) and the reality of death (*réel de la mort*), respectively inspired by the theories of Freud (1895), Janet (1889, 1904) and Lacan (1973).

François Lebigot and l'effroi de la mort

In this classically oriented French theoretical approach of stress and trauma, the core of a potentially traumatizing event is the confrontation with the 'reality of death' (*le reel de la mort*). Due to the sudden and unexpected confrontation with death, the subject enters into a temporary state of petrified shock – i.e. seems to turn into a stone as numerous soldiers in the old Greek Medusa tale (Crocq, 1999). In this state of *effroi de la mort* (*dread of death or the frozen state of shock in the face of death*), the subject encounters the complete and absolute *emptiness* and *nothingness of death*, and experiences it as being totally abandoned by the world of the living.

This interpretation is well-documented by numerous clinical vignettes in the work of Lebigot (2005). The notion of *effroi de la mort* is essential in this approach: it describes the impact of an event which involves extreme anxiety, horror, total abandonment and powerlessness, which also appears in DSM-IV under the A2 criterion of PTSD (APA, 2000) but which has since been dropped in DSM-5 on the basis of statistical studies of its predictive power in assessing PTSD. This classical trauma interpretation is inspired by the early Freudian viewpoints on trauma in which the psyche is represented by a restricted volume in which there is a network of representations, between which small amounts of energy (affect) can freely circulate (cf. Figure 1). This restricted volume is surrounded by a membrane which is seen as a stimulus barrier, protecting the psyche from being overwhelmed by over-stimulation and being charged with positive energy. This metaphorical view on the psyche is used to explain the difference between stress and trauma. The stimulus barrier protects the psyche from excessive stimulation which could, from the outside, influence the psychological equilibrium on the inside and deform the shape of the stimulus barrier. The tension on the outside of the stimulus barrier can be seen as a kind of strain ('to stress' literally means 'to increase the strain') which will have a negative impact on the free circulation of energy between the representations within the psychological system. The anxiety which results from this specific state will lead to an increase of energy on the outside of the stimulus barrier (i.e. the subject will draw on its own resources and react). The psyche defends itself against the aggression coming from the outside. When the tension from the outside decreases or disappears, the stimulus barrier takes its previous shape again. In this interpretation, the subject defends itself against a threat from the outside and manages to recover from stress: an apt metaphor for psychological resilience.

In the case of a traumatic impact, the story becomes different. Trauma is seen as the vital threat, life-threatening and striking unexpectedly and very suddenly. It is an overwhelming amount of energy coming from the outside which abruptly penetrates the stimulus barrier. It infiltrates into the network of representations (cf. Figure 2) and discharges an enormous amount of energy within it which instantaneously disrupts the psychological equilibrium. The external aggression leaves a strange nucleus – *corps étranger* – in place i.e. an element which has no representation, no symbolic shape and which can therefore not be integrated in the existing network of mental representations. According to this dynamic view, this strange corpus is related to the close encounter with the 'reality of death' and cannot be 'metabolized' because there is no representation, no symbolisation, for death in the broader network of representations. The traumatizing experience literally stops time in the psyche and leaves the subject in a state of total abandonment.

In this metaphoric interpretation, the psyche is considered as a restricted volume surrounding a network of representations, separated from the outside world by a membrane. This view illustrates the difference between vital – i.e. threats for the integrity - and non-vital aggressions on the psyche. The way in which the psyche remains affected by a (potentially) vital aggression from the outside and/or tries to recover from it, as a function of time, is determined by the possibilities of symbolizing or representing the experience. The inability to symbolize the traumatic experience through verbal expression and language excludes the traumatized subject from the world of those who are able to speak and renders integration within the network of mental representations impossible. Social relations and all of the internal and external networks based on that world of internal representations are profoundly disrupted resulting in loss of the ability to relate to self or others.

Stress involves a psychological suffering which persists for as long as the external pressure (strain) lasts; when the 'stress' disappears, the membrane or stimulus barrier slowly takes back its previous shape and the system recovers. The internal functioning of the psyche can continue, even in the light of small internal changes which might have occurred due to the losses provoked by the external aggression. On the other hand, trauma leaves – potentially forever – an internal source of disruption, an indigestible strange corpus (*corps étranger*) within the psyche; this disruption will persist, even when the threat is over. But stress and trauma often go hand in hand: in these cases, there is first the crushing and transforming impact due to the strain from the outside of the membrane, and then the penetration by the strange, disrupting element. In the case of trauma, the protective function of the stimulus barrier fails and is penetrated: the psychic apparatus becomes the subject of an *effraction*, which literally means break-in or brutal impact. Once penetrated, the psychic apparatus will do everything it can to assimilate or reject the new source of excessive stimuli but these attempts fail. It appears impossible to metabolize the pathogenic nucleus of trauma. On the clinical level, first symptoms of severe stress will appear before the appearance of the *repetition syndrome* (*syndrome de répétition*) which may begin much later. In this syndrome, there will always be the re-appearance of the strange nucleus (*corps étranger*) under its various forms, like a testimony of the *effraction* of the psychic apparatus.

As Louis Crocq describes in Les traumatismes psychiques de guerre (1999): '*Raymond F., a 17 year old soldier in the Wehrmacht in 1944, was suddenly confronted with a Russian soldier who was lying severely wounded on the side of a road, with an open abdomen. In order to liberate this enemy soldier from his suffering, the German officer achieved him with a single shot in the head. The last thing the poor soldier did, was looking Raymond right into the eyes. Ever since then, each night when trying to fall asleep, Raymond relives this scene, over and over again, in greatest detail, as if he were able to count the grassroots on which the Russian soldier was lying* (p.98)'.

The hermeneutic aspect of this metaphorical trauma view, which is proposed by contemporary French authors (Barrois, 1998; Crocq, 1999, 2001; Lebigot, 2001, 2005), indicates that a potentially traumatizing event, simply called *le trauma* in French, is experienced subjectively. But what is traumatizing for one survivor, will not necessarily be traumatic for the other. The essence of trauma has to be seen as an encounter between an event and a subject. From this angle of incidence, each event could be seen as potentially traumatizing, but a subject with sufficient defensive resources (i.e. a strong enough stimulus barrier) will be able to formulate an adapted response to it.

The specificity of traumatization: From the close encounter with death to the repetition syndrome (*syndrome de répétition*).

Barrois (1998) presented psychological trauma as a disruption in the connection with the external world. The subject is confronted with *the unthinkable*, namely his own death. This confrontation, for which one can never be fully prepared, triggers a reaction of *effroi* (frozen fright or dread) in the individual, in which *death of the self* is announced as the ultimate truth. Trauma is experienced as a disruption in the continuity of one's existence. And even if we all know that death is the inescapable end for every human being, we do not have a (mental) representation of death in our psyche. We live our lives as if we all know that we will die, but without believing that this will actually happen. We all know what a dead body is, but we do not know what death in itself is really about. The real cognition of death (*I will die*) suddenly appears as a consequence of an intimate encounter with death; our own death or the death of another human being who matters to us.

According to this theoretical approach, the psychological trauma response will be an answer, a syndrome of adaptation, to the disruption of the continuity of life: there will also be a disruption between the life prior to the trauma, on the one hand, and the post-trauma life, on the other hand. But in the light of a potentially traumatizing event, not everyone will be traumatized. If the subject can resist/prove resilient and find a way to cope with the confrontation with death and complete the necessary series of survival reactions in the face of the confrontation with the reality of death, the neurobiological symptoms will dissolve and disappear. In the psyche, there will still be a conscious memory of the potentially traumatizing event, but this memory will become less intense and more vague over time. The opposite is also true: when there is no fulfilment or completion of coping with the potentially traumatic event, there will be no final narrative to insert the traumatic memory into the personal life story and psychotraumatic symptoms will tend to coalesce and persist.

As already mentioned, the concept of a repetition syndrome through which the traumatic experience, containing the 'irrepresentable nucleus' in the original event, reappears over and over again, plays an essential role in the work of contemporary French authors (Andreoli & Damsa, 2005: Barrois, 1998; Crocq, 1999, 2001; Lebigot, 2001, 2005). Symptoms of intrusion are described as repeated involuntary reminiscences with distinct manifestations, expression registers and circumstantial appearances (Crocq, 1999). Beyond the automatic replication of the experience of frozen fright (*effroi*) and petrification (*sidération*), this approach also highlights the memory and personality disturbances which often develop after some time (cf. the delayed onset of symptoms as described in DSM) and involves psychological functions which are both complex and independent from the direct effects of overwhelming stress (Crocq, 1999; Crocq, 2002; Lebigot, 2002).

Through the repetition syndrome, these authors describe how the traumatizing event instantly stops or freezes the time in the psyche and installs itself, instantly or after a given amount of time, as a strange object which creates a permanent reiteration of the traumatic experience, accompanied by exactly the same somatosensory and psychological phenomena as those present at the time of the original event. The nature of the alterations in memory and consciousness occurring during the repetition syndrome have been described by Crocq (1999, 2001), and Crocq and De Verbizier (1989) with the use of the concepts of 'traumatic unconscious (*inconscient traumatique*)', dissociation from consciousness and catharsis.

The 'black hole' of the traumatic memory is considered as the characteristic trait of posttraumatic pathology. The French authors describe the concept of 'traumatic unconscious' as a result from the close encounter with the reality of death (*'réel de la mort'*). The experience of perplexity, absurdity and annihilation, which typically appears in clinical descriptions of trauma survivors, constitutes a 'short circuit' of symbolic function i.e. the cognitive action responsible for the symbolization of experience is suspended, rendering the subject without words or mental representation of the experience. In this view, psychological traumatisation is brought back to disruption in 'signifiers' – i.e. the Lacanian concept for meaningful representation carriers, words, which are needed to structure knowledge – due to the sudden confrontation with the unthinkable and the inconceivable. In line with this view, trauma means a disruption between the 'signified' (i.e. death) and the 'signifier' (i.e. the representation or symbolisation of death in the network of representations in the psyche). What really happens is that the previously symbolic references to death prove insufficient in the face of the *réel de la mort* which then leaves a break between the signified and the signifier. Thus, the psychological disturbances occurring in the repetition syndrome are not the direct consequences of the stress created by the apperception of an event, but of the psychic impact of this same apperception and its total inconceivable (irrepresentable) or unthinkable aspect. It should be noted that the '*réel de la mort*' does not exactly equate with the 'reality of death' even if that is the closest literal translation available in English. *Réel de la mort* more specifically refers to a fundamental confrontation with the 'wordless shadow of life', the 'nothingness', the 'hole' left by trauma in the current psychological functioning of the stricken subject and the annihilating effect which will continue over time, upon each automatic reiteration of this experience through the repetition compulsion.

These metaphoric French trauma interpretations, influenced by the terminology and the clinical concepts of Freudian and Lacanian psychoanalytic theories, lead to a more phenomenological approach which is relevant for the interpretation of trauma-related accounts of survivors. They illustrate the specific meanings and position of stress and trauma concepts in the French-oriented trauma literature as they are widely known in Latin countries (i.e. France, Spain, Italy, Portugal and the South-American countries). Clinically, these interpretations may allow us to make a better sense of the accounts of trauma survivors who always repeat the fact that their experience is impossible to express in words while clinicians are always in search for the right therapy protocol or technique which make the expression of trauma-related experiences possible.

Discussion and conclusion

The classical French authors can be credited with an important descriptive clinical contribution to the field of psychological trauma. This article explored how contemporary French authors, inspired by classical psychodynamic theories, have contributed to a more integrative view on a broad range of terms and concepts used by trauma practitioners in different language groups. A problem with the French terminology is that *le trauma* (*the trauma*) is used both as a trauma-eliciting event and as a chronic state of psychological trauma thus conflating cause and origin. Furthermore, these theoretical viewpoints are clinically interesting but difficult to confirm or to falsify through empirical research. This may be the primary reason why the current mainstream Anglo-Saxon literature, which is essentially driven by empirical research rather than by clinical experience, has not taken these insights into account till now.

French theory separates the clinical pattern of psychological trauma on two distinct levels: the level of posttraumatic symptoms and the level of the underlying personality which creates and sustains the explicit symptoms. The lack of filtration of stimuli, the loss of interest in former activities and emotional withdrawal are in fact only expressions of this alteration of the personality in the wake of a traumatizing event. Instead of focusing on the biological, behavioural and cognitive aspects of psychological trauma, French theories focus on the phenomenological variety and the underlying meaning of trauma symptoms.

One of the weak points in the mainstream anglophone literature is its considerable variation in terminology, while there is a huge amount of research on the determinants and predictors of (both neurobiological and psychological) pre-, peri- and posttraumatic symptoms. The translation of the empirical results into relevant clinical concepts or therapeutic approaches seems sometimes difficult. In DSM-IV, DSM-IV-TR and DSM-5 (APA, 1994, 2000, 2013), PTSD is essentially approached from a quantitative point of view as defined by a set of symptoms, belonging to three different clusters, and by time-related criteria of dysfunction. A possible bridge toward the French clinical trauma interpretations could be perceived in the split of the *stressor criterion* of PTSD into two parts (in DSM-IV-TR); Criterion A1 and Criterion A2. Criterion A1 refers to the objective elements of the trauma (e.g. physical threat of the event to self and others), and Criterion A2 relates to subjective components (e.g. the individual's response): intense fear, helplessness, and/or horror (APA, 1994; 2000) yet this potential crosswalk between theoretical approaches has been lost following the redefinition of the Stressor Criterion in DSM-5. The subjective aspect of the trauma diagnosis, focusing on the meaning that a potentially event had for its survivor, could receive even more attention since recovering from trauma should not only be reduced to showing less symptoms or a reduction in the intensity (severity) of the trauma symptoms listed in the typical symptom clusters. The phenomenology of trauma, as it appears in French contemporary trauma theories, insists on the adaptive ability to reach a level of integration of the traumatic experience into one's psychic existence by transforming the 'traumatizing experience' into a narrative, accompanied by a catharsis of affects and the cessation of the compulsive repetition syndrome ('*syndrome de répétition*'). This is in line with the opinion of Van der Hart, Nijenhuis & Steele (2006) who consider failed integration on the perceptual, cognitive and somatic levels as essential features of psychological trauma and of Krystal (1988) who described psychological trauma in terms of alexythymia.

French trauma theories lead to a more psychodynamic and hermeneutic trauma interpretation which is relevant from the clinical point of view – i.e. closely matches the accounts of trauma survivors – and offers insights which go beyond the descriptive approach of psychological trauma in the diagnostic nomenclature.

It would be meaningful to integrate both approaches within a more holistic trauma view and bridge the gap between the quantitative (empirical) trauma approach, on the one hand, and more qualitative (clinical) interpretations, on the other hand.

There should also be a search for a more restrictive terminology in the field of psychotraumatology, aiming at what psychological trauma really is: a close encounter with the inconceivable, the irrepresentable, which leads to a loss of integration of the various levels of the personality.

DSM-5 (APA, 2013) lists symptoms which are essentially applicable in the stress and HPA-activation paradigm, while the persisting symptoms of trauma survivors appear to be of psychodynamic origin. In the light of French theory, the following question might be raised: Should the new stressor criterion of DSM V, i.e. the exposure to actual or threatened death, serious injury or sexual violation, remain in the definition of PTSD? The current nosography of posttraumatic stress disorder suffers from a lack of conceptual richness or of words which fit with the actual experience of trauma survivors. While the French interpretation of psychological trauma might be the unspeakable experience of the confrontation with the 'reality of death', a more nuanced view might be the confrontation with an aspect of life which one cannot fully integrate, associated with manifest neurobiological correlates and a typical constellation of stress and anxiety symptoms. Therefore, from a classical hermeneutical and clinical point of view, it would make more sense to broaden the A2 criterion for PTSD instead of eliminating it from the set of diagnostic criteria.

The scientific evidence in neurosciences and behavioural sciences which have appeared over the past two decades, should make it possible to bridge the phenomenological description of trauma in French theory with DSM-5 criteria: intrusion symptoms, persistent avoidance, negative alternations in cognitions and mood, alteration in arousal and reactivity, duration of these symptoms and the distress or impairment in social, occupational, or other important areas of functioning. The increase of cross cultural and 'cross conceptual' research, in a context of global integration of trauma societies, will hopefully lead to the development and further improvement of evidence-based clinical treatment for trauma survivors which should show more respect for cultural and transglobal diversity.

Acknowledgements

The authors would like to thank Prof. dr. Onno van der Hart from the Faculty of Social Sciences of the Utrecht University for his careful and critical reviews of the previous versions of this article and Prof. dr. Harold Stephen Kudler from the VA Medical Center (Durham, USA) for his valuable in-depth reading of this article which lead to additional insights and adaptations..

Authors' contributions

EDS conceptualized this article on the basis of his knowledge of the current French theories on psychotrauma, he drafted the manuscript of this article and collected relevant literature references. RK and JM critically reviewed the document for intellectual content and style, and have given final approval for this version to be published.

References

American Psychiatric Association (1980). *Diagnostic and statistical manual of mental disorders* (3rd ed.). Washington DC: Author.

American Psychiatric Association (1987). *Diagnostic and statistical manual of mental disorders* (3rd ed. Rev.). Washington DC: Author.

American Psychiatric Association (1994). *Diagnostic and statistical manual of mental disorders* (4th ed.). Washington DC: Author.

American Psychiatric Association (2000). *Diagnostic and statistical manual of mental disorders* (4th ed. Text editon). Washington DC: Author.

American Psychiatric Association (2013). *Diagnostic and statistical manual of mental disorders* (5th ed.). Washington DC: Author.

Andreoli, A., & Damsa, C. (2005) Stress, dissociation ou trauma psychique? Incidence des travaux francophones dans le débat international. *Revue francophone du stress et du trauma, 5(2)*,61-65.

Barrois, C. (1998). *Les névroses traumatiques.* Paris: Editions Dunod.

Belenky, G. (1987). *Contemporary studies in combat psychiatry.* Westport, CT: Greenwood Press.

Bramsen, I. (1995). *The longterm psychological adjustment of World War II survivors in the Netherlands.* Delft: Eburon Press.

Brewin, C.R., Andrews, B., & Valentine, J.D. (2000). Meta-Analysis of Risk Factors for Posttraumatic Stress Disorder in Trauma-Exposed Adults. *Journal of Consulting and Clinical Psychology,* 68,748-766.

Breslau, N., & Davis, G. C. (1992). Posttraumatic stress disorder in an urban population of young adults: Risk factors for chronicity. *Journal of Psychiatry, 149,* 671-675.

Breuer, J., & Freud, S. (1973). *Studies on Hysteria: Physical Mechanisms of Hysteria. Translated by Strachey 1.* New York: Penguin.

Carlson, E. B., Furby, L., Armstrong, L., Shlaes, J. (1997). A conceptual framework for the long-term psychological effects of traumatic childhood abuse. *Child Maltreatment, 2(3),* 272-295.

Charcot, J.-M. (1887). *Leçons sur les maladies du système nerveux faites à la Salpêtrière.* Paris : Progrèm Médical Delahaye & Lecrosnie.Charpentier, R. (1917). Allocution de M. René Charpentier. *Annales médico-psychologiques, 74(8),*273-329.

Crocq, L, & De Verbizier, J. (1989). Le traumatisme psychologique dans l'oeuvre de Pierre Janet. *Annales médico-psychologiques,147,*983-987.

Crocq, L. (1992). Le syndrome de répétition dans la névrose traumatique. Ses variations cliniques, sa signification. *Perspectives Psy, 32(2),*59-65.

Crocq, L. (1999). *Les traumatismes psychiques de guerre.* Paris: Odile Jacob.

Crocq, L. (2001). Le traumatisme psychique dans la pensée psychiatrique francophone. Introduction. In: De Clercq, M., & Lebigot, F. (Eds.) *Les traumatismes psychiques.* Paris: Masson.

Crocq, L. (2002). Persée, la Méduse et l'effroi. *Revue francophone du stress et du trauma, 2(3),*133-138.

De Clercq, M., & Lebigot, F. (2001). *Les traumatismes psychiques.* Paris: Masson.

Cygielstrejch, A. (1916). La psychologie de la panique pendant la guerre. *Annales médico-psychologiques,7,* 172-92.

Delcasse, J.A. (1834). *De la commotion.* Paris.

Dowbiggin, I. (1985). Degeneration and Hereditarianism in French Mental Medicine, 1840-1890 : Psychiatric Theory as Ideological Adaptation. In: Bynum, W.F., Porter, R., & Shepherd, M. (Eds.). *The Anatomy of Madness: Essays in the History of Psychiatry,* Vol. 1, 188-232. London: Tavistock.

Erichsen, J.E. (1875). *On concussion of the spine.* London: William Wood.

Fenichel, O. (1945). *The psychoanalytic theory of neurosis.* New York: Norton & Company Inc.

Ferenczi, S., Abraham, K., Simmel, E.& Jones, E. (1921). Psychoanalysis and the war neuroses. *The International Psycho-Analytical Library, 2,*1-59. London, Vienna, New York: The International Psycho-Analytic Press.

Foa, E.B., Rothbaum, B.O. (1998). *Treating the trauma of rape.* New York: Guilford Press.

Foa, E. B., Riggs, D. S., & Gershuny, B. S. (1995). Arousal, numbing, and intrusion: Symptom structure of PTSD following assault. *American Journal of Psychiatry, 152,*116-120.

Freud, S. (1893/2001). Some points for a comparative study of organic and hysterical paralyses. In J. Strachey & A. Strachey (Eds.), *Standard edition of the complete Psychological Works of Sigmund Freud,* 1. London: Hogarth.

Freud, S., & Breuer, J. (1893). Preliminary Communication. In: *Studies on Hysteria,* Standard Edition, 2:17.

Freud, S. (1918). *Reflections on war and death.* New York: Moffat, Yard & Co.

Freud, S. & Breuer, J. (1956). *Etudes sur l'hystérie* (7th ed., 1981). Paris: PUF.

Gersons, B.P.R., Carlier, I.V.E. (1992). Post-Traumatic Stress Disorder: The History of a Recent Concept. *British Journal of Psychiatry, 161,*742-748.

Horowitz, M. (1986). Stress-response syndromes: A review of posttraumatic and adjustment disorder. *Hospital and Community Psychiatry, 37,*241-249.

Janet, P. (1889). *L'automatisme psychologique.* Paris: Alcan.

Janet, P. (1904). L'amnésie et la dissociation des souvenirs par l'émotion. *Journal de psychologie, 1,*417-453.

Janoff-Bulman, R. (1985). The aftermath of victimization: Rebuilding shattered assumptions; In: Figley, C.R. (Ed.), *Trauma and its wake: The study and Treatment of Post-Traumatic Stress Disorder,* NewYork: Brunner-Mazel.

Jones, D.W. (1982). The Vietnams of Michael Herr and Tim O'Brien: Tales of Disintegration and Integration. *Canadian Review of American Studies, 13(3),*309-200.

Kardiner, A., & Spiegel, H. (1947). *War stress and neurotic illness.* New York-London: Paul B. Hoeber Inc.

Kinzie, J.D., Goetz, R.R. (1996). A Century of Controversy Surrounding Posttraumatic Stress-Spectrum Syndromes: The Impact on DSM-III and DSM-IV. *Journal of Traumatic Stress, 9(2),*159-179.

Kleber, R. J., & Brom, D. (1992). *Coping with trauma.* Amsterdam: Swets & Zeitlinger.

Kraepelin, E. (1899). *Ein Lehrbuch für Studierende und Aertze,* 2 vol., Leipzig: Johann Ambrosius Barth.

Lacan, J. (1973). *Les quatre concepts fondamentaux de la psychanalyse.* Paris: Seuil.

Lefebvre, P.,& Barbes, S. (184). L'Hystérie de guerre : étude comparative de ses manifestations au cours de deux derniers conflits mondiaux. *Annales médico-psychologiques, 142 (February),* 262-266.

Krystal, H. (1988). *Integration and Self Healing : Affect – Trauma – Alexithymia.* New York : Routledge.

Lebigot, F. (2001). Prises en charge psychothérapeutiques psychodynamiques. In: De Clercq, M. & Lebigot (Eds). *Les traumatismes psychiques.* Paris: Masson.

Lebigot, F. (2002). L'effroi du traumatisme psychique. Le regarder en face ou s'en protéger. *Revue francophone du stress et du trauma, 2(3)*,139-146.

Lebigot, F. (2005). *Traiter les traumatismes psychiques. Cliniques et prise en charge.* Paris: Editions Dunod.

Lindy, J., Green, B., & Grace, M. (1987). The stressor criterion and posttraumatic stress disorder. *The Journal of Nervous and Mental Disease, 175*,269-272.

Micale, M.S. (1985). The Salpêtrière in the Age of Charcot : An Institutional Perspective on Medical History in the Late Nineteenth Century. *Journal of Contemporary History, 20* (October), 703-731.

Micale, M.S. (1996). Paradigm and Ideology in Psychiatric History Writing: The Case of Psychoanalysis. *Journal of Nervous and Mental Disease, 184*, 146-52.

Micale, M.S. & Lerner, P. (2001) (Eds). *Traumatic Pasts. History, Psychiatry, and Trauma in the Modern Age, 1870-1930.* New York: Cambridge University Press.

Pinel, P. (1798). *Nosographie philosophique, ou la method de l'analyse appliquée à la Médecine.* Paris : Richard, Caille, & Ravier.

Roussy, G., & Lhermitte, J. (1919). *Les psychonévroses de guerre.* Paris : Masson.

Van der Hart, O., & Friedman, B. (1989). A Reader's Guide to Pierre Janet on Dissociation: A Neglected Intellectual Heritage, *Dissociation, 2*, 3-16.

Van der Hart, O., Brown, P., & Van der Kolk, B.A. (1989). Pierre Janet's Treatment of Post-Traumatic Stress, *Journal of Traumatic Stress, 2*, 379-395.

Van der Hart, O., Nijenhuis, E., Steele, K. (2006). *The haunted Self: Structural dissociation and the treatment of chronic traumatization.* New York: Norton.

Wilson, J.P., Harel, Z., & Kahana, B. (1988) (Eds). *Human Adaptation to Extreme Stress From the Holocaust to Vietnam.* New York: Springer Science.

Wilson, J.P. (1989). A person-environment approach to traumatic stress reactions. In: J. Wilson (Ed.) *Trauma, transformation and healing: An integrative approach to theory, research, and post-traumatic therapy*, 75-110. New York: Brunner/Mazel.

Wilson, J.P. (1994). The Historical Evolution of PTSD Diagnostic Criteria: From Freud to DSM-IV. *Journal of Traumatic Stress, 7(4)*, 1994.

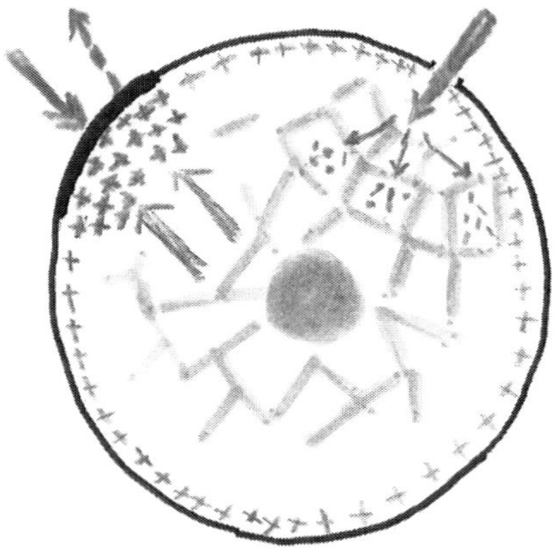

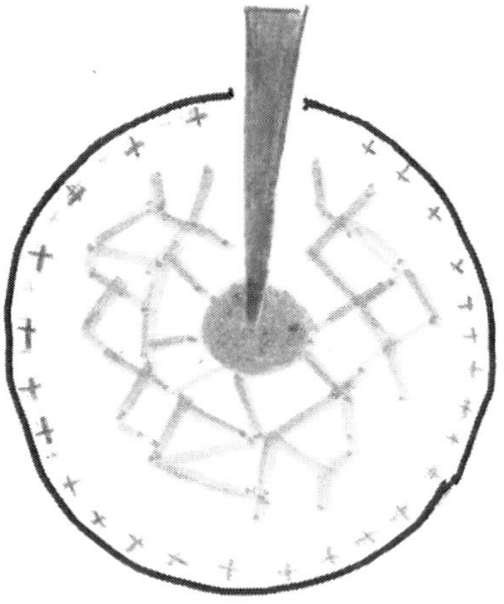

Figure 1 shows how the stimulus barrier is reinforced and protects the psyche from too much stimulation; it filters and fragments sensorial inputs, allowing them to be assimilated in the network of representations

Figure 2 shows how a stimulus with maximum excitation breaks into the stimulus barrier and penetrates deeply into (non-verbal part of) the network of representations where it finds no signifying link

CHAPTER 4

A Phenomenological Analysis of Disaster-Related Experiences in Fire and Emergency Medical Services Personnel

E. De Soir, M. Knarren, E. Zech, J. Mylle, R. Kleber & O. van der Hart

Paper published in *Prehospital and Disaster Medicine* 2012;27(2):1-8
Adapted version published in French in *Revue Francophone du Stress et du Trauma* – Juin 2011 – Tome 11 – N°2

A Phenomenological Analysis of Disaster-Related Experiences in Fire and Emergency Medical Services Personnel

Erik De Soir, Marcia Knarren, Emmanuelle Zech, Jacques Mylle, Rolf Kleber & Onno van der Hart

Summary

This chapter explores the experiences in fire and emergency medical services personnel during and immediately after a technological disaster using a phenomenological approach. Personnel engaged in the rescue operations during and immediately after the Ghislenghien gas explosion reflected upon their experiences in their responses to a specially designed, self-reporting questionnaire. The questionaire also contained some open ended questions. Firefighters reported more perceived threat and direct exposure to death than did emergency medical services personnel. Qualitative analysis indicates that the central characteristics of this potentially traumatizing event were the suddenness and massiveness of the impact and the fact that it involved young victims and/or multiple deaths. With regard to emotions, powerlessness, horror, fear, a sense of apocalypse, and grief were experienced by both firefighters and emergency medical services personnel. As to the most shocking aspects of their disaster experience, fire fighters noted the death of colleagues, the involvement of friends and family, the massive impact, and exposure to the burned victims as most shocking. Emergency medical services personnel and in-hospital staff reported the impact, the confrontation with death, the involvement of friends and family, and the pain, suffering, and screaming of burned victims as the most shocking aspects of this event. Qualitative differences in the lived experiences of firefighters, emergency medical services personnel, and in-hospital staff might be explained by differences in life threat, contact with death, and various degrees of training.

Keywords: disaster experiences, emotions, characteristics, fire & rescue services

Introduction

Individuals exposed to stressful, potentially traumatizing events are at risk for developing post-traumatic stress reactions (PTSR), acute stress disorder (ASD), and/or post-traumatic stress disorder (PTSD). Research on the effects of such events usually concentrates on primary victims, i.e., those directly threatened or injured in an event (Dirkzwager, Grievink, Van der Velden & Yzermans, 2006; Soeterman et al., 2006). Less is known about the specific contextual variables that might lead to PTSR in high-risk occupational groups such as firefighters, rescue services, and emergency personnel (Beaton & Murphy, 1995; Durham, McCammon & Allison, 1985; Morren, Yzermans, Van Nispen & Wevers, 2005). Understanding these variables might lead to more adequate psychosocial support for fire and emergency medical services personnel.

The weak point of most studies of occupational trauma in emergency workers is that the research has focused on event-specific (i.e., objective) aspects instead of the lived experiences of potentially traumatizing rescue interventions, which indicate the way in which these events are experienced on a subjective level. In times of disaster, these target groups are more specifically at risk for encountering traumatizing events (De Soir, 1999; Fullerton & Ursano, 1997; Laposa & Alden, 2003; Leon, 2004; Mitchell & Dyregrov, 1993; Smith & Roberts, 2003; Ursano, Fullerton, Vance & Kao, 1999; Wagner, Heinrichs & Ehlert, 1998; Weiss, Marmar, Metzler & Ronfeldt, 1995). Indeed, they are working with the injured or the dead, and therefore are exposed to a variety of stressors. Ambulance personnel working in prehospital care often are exposed to stressful events in their daily routine – but, large-scale events or disaster situations add specific stressors to the routine rescue work. As a result, ambulance personnel are at risk for developing a PTSD after a traumatic event (Clohessy & Ehlers, 1999). Overall, it is to be expected that rescue work has detrimental emotional and psychological effects on the well-being of fire and emergency services personnel.

In fact the incidence of PTSD after experiencing a traumatic event in the general population is 10 to 15% and lifetime prevalence for PTSD is estimated to be 7.8% (Kessler, Sonnega, Bromet & Hugher, 1995); however, estimates for specific at-risk populations are higher. For example, the prevalence of PTSD in German firefighters was as high as 18.2% (Wagner, Heinrichs & Ehlert, 1998). According to Clohessy and Ehlers (1999) as well as Grevin (1996), estimates of the prevalence of PTSD in emergency personnel approximately 20%. However, others reported a PTSD prevalence of 3 to 7% in rescue workers (Hodgkinson & Stewart, 1991). These differences in the prevalence of PTSD might be related to both situational event-specific and person-related factors such as trauma exposure, but this is unclear (Myles, Levine, Ramsden & Swanson, 1990). While most studies seem to indicate levels of posttraumatic stress based on PTSD prevalence, no study has investigated the qualitative aspects of the psychological post-intervention sequelae in fire and emergency services personnel during and after disasters. Such information could broaden our understanding of the lived experience of potentially traumatizing events of fire and emergency services during disasters in order to allow better psychosocial support for these target groups.

One study actually analyzed the written stories from 52 ambulance nurses and emergency medical services technicians who each described a traumatic event (Jonsson & Segestin, 2003). Those descriptions were analyzed using Van Kaam's method (Van Kaam, 1959; Van Kaam, 1966; Van Kaam, 1969). Their findings indicate that the nurses and technicians showing a strong identification with the victims experienced more overwhelming emotions and feelings than did those who did not identify so strongly. Identification with the victim is a strong predictor of post-traumatic stress among helpers (Ursano, Fullerton, Vance & Kao, 1999), especially when children are involved (Dyregrov, 1992; Dyregrov, 1993). In addition, several studies (Green, Grace & Gleser, 1985; Hartsough, 1985; Martelli, Waters & Martelli, 1989) have highlighted a series of situational risk factors for the development of post-traumatic stress symptoms among emergency services personnel: (1) the victims' ages (e.g., injuries/deaths in infants and children have a greater impact than do adults); (2) exposure to gruesome injuries and/or death; and (3) facing dangerous and/or unpredictable situations.

The cognitions and emotions experienced during an event also are predictors of post-traumatic distress (Jeavons, Greenwood & Horne, 2000). According to Herman (2002), when encountering a traumatic event, intense fear, helplessness, powerlessness and threat of death are common. However, from a resilience perspective, in which the focus is on the personal strengths of people, it can be expected that some emergency responders may report no complaints or symptoms related to the traumatic stress experience, and sometimes may even report on growth instead of distress. There exists extensive research on the positive consequences after confrontation with adversity, emotionally disturbing and potentially traumatizing events (Bravo, Rubio-Stipec, Canino, Woodbury & Ribera, 1990; Shakespeare-Finch, Smith & Gow, 2003; Tugade, Frederickson & Barrett, 2004). These positive effects are related to positive affects and cognitive processing of the traumatizing event resulting in, for example, a perception of decreased experienced threat, lower risk for physical injury, and less uncontrollability of the situation. The positive consequences of task-oriented incidents (i.e., fighting fires) and person-oriented incidents (i.e., rescue work) should be distinguished (Moran, 1998). In this study, firefighters who contributed to the rescue of disaster victims described less positive reactions in their rescue experience than had those involved only in fighting the fire. Thus, it is expected that the positive impact of emergency work should occur more often with task-focused activities than among rescue personnel involved in more person-focused incidents.

In the current study, the experiences of the personnel who served at the disaster scene or in the regional hospitals after the massive gas pipe explosion that occurred in an industrial park in Ghislenghien, Belgium, on 30 July 2004 were explored. Debris from the gas pipe weighting several tons was thrown as high as 200 meters into the air. The heat of the fire was felt at nearly 2 kilometers from the explosion site. Debris from buildings was projected up to 6 kilometers. In total, the explosion killed 24 people; mainly factory workers, firefighters, and a police officer, and about 132 people were wounded – many suffered severe burn injuries. Most of the wounded victims were factory workers, firefighters, police officers, and car occupants passing by at the moment of the explosion.

The similarities and differences between the experiences of firefighters and emergency medical personnel (including in-hospital rescue staff) are examined by analyzing the reported emotions and cognitive reactions to the event at 4 months (T1) and 14 months (T2) after the event. Since the data collected on the exposure among firefighters and emergency medical personnel did not allow a *pairwise* comparison at T1 and T2, detailed comparison of the evolution over time were not included in the aims of this study. The characteristics of the event that were most reported were investigated before the assessment of the feelings and emotions. A qualitative analysis method of the answers to open-ended questions was used to assess which aspects were viewed as the most shocking. Similarly, to Laposa and Alden (2003), it was expected that firefighters would report their work-related experiences were more dangerous physically and involved a greater threat to their lives and highlighted the death of five fire colleagues during the event, compared to those of emergency medical services and hospital staff personnel. Since the emergency medical services and in-hospital staff personnel were exposed more directly to human suffering, but less confronted with life threat, it was expected that they would report more on the emotional burden of this unusual work. These hypotheses lead to different expectations regarding the type of potentially traumatizing events to which both target groups were exposed. It was expected that emergency services personnel and in-hospital emergency responders would describe more victim-related experiences and that firefighters would report more exposure to the event.

Method

Participants

This study is part of a broader scientific research project led by the Stress and Trauma Research Center of the Royal Military Academy, the Faculty of Psychology of the University of Louvain, the Service d'Appui Psychologique aux Intervenants (Psychological Support Service for Caregivers) and the Union Royale des Sapeurs-Pompiers du Hainaut (Royal Association of Firefighters of the Province of Hainaut). Only fire, rescue, and emergency medical personnel who participated in the rescue operations on the scene of the event in Ghislenghien or in the emergency management in the regional hospitals were approached. They were contacted through their normal command structure. Two self-report questionnaires were distributed in a closed envelope at T1 and T2, and were returned in an anonymous way; the closed envelopes were collected at the fire stations or hospitals and were collected by personal contact.

Finally, 103 firefighters and 77 emergency medical personnel filled in the questionnaires at T1; the response rates were respectively 42.0% and 31.4%. The group of fire fighters consisted of 100 men and three women, aged between 18 and 59 years of age (mean = 40 ±10.7 years). The emergency medical group consisted of 22 men and 55 women, between 24 and 57 years old (mean = 40 ±9.1 years). Only 23 male fire fighters and 35 emergency medical personnel (12 men and 23 women) filled in the questionnaires at T2. Their age ranged from 23 to 58 years (mean = 39 ±10.2 years) and 23 to 55 years (mean = 40 ±9.2 years) respectively. The procedure used for distributing the questionnaires did not allow the use of registration numbers to keep track of the participants from T1 to T2. There is only anecdotic evidence explaining the important loss of participants between the measurements at T1 and T2; i.e. 75% among the fire fighters (from 103 to 22) and 50% (from 77 to 35) among the emergency personnel (see limitations).

Measures

Description of the Instruments—To assess the experiences of the above-mentioned population, a questionnaire was designed to fit the Ghislenghien disaster. It contained: (1) an informed consent; (2) a set of demographic questions; (3) a section related to the experiences on the scene of the accident (the epicenter) or in the periphery; (4) questions related to the current emotional experiences; and (5) questions related to both available social support and professional help. Additionally, one question was used to assess whether there were elements about which the respondent did not want to discuss and the reasons, and finally, a blank page for comments.

Open-ended questions addressed: (1) how respondents experienced the event and its aftermath ("Please describe how you experienced the Ghislenghien disaster and how you got involved in it; e.g., "When did you arrive, where were you, what did you precisely do or see, what did you feel?"); (2) the most shocking aspect of the event ("Which aspect of this event shocked you most e.g. of what you heard, smelled, certain behaviors, a memory, a person, an object)? Describe this as precisely as possible."; (3a and 3b) additional professional or private aspects having influenced their reactions ("Are there in your professional/private environment additional things (e.g., reactions and/or behaviors, or the organization itself) that touched or shocked you at the moment of the event or as a consequence of it?". The following was written at the top of the blank page: "This space has been entirely saved for you in order to allow you to express yourself freely, without limitations, about what happened to you (in the context of the event), about the way you lived through the event and about the way that you still deal with it today. You can also write on the impact that the event had on you, your life and your family. You can also write about the way that you deal with this event today, about the help that you get from other people or about the help you think you need in order to recover from the event. If you need more space, do not hesitate to add some more paper." Approximately 30 minutes were required to complete the questionnaire.

Qualitative Data Analysis

In order to focus on the qualitative aspects of the experiences of the fire and emergency services personnel involved in the Ghislenghien event, a phenomenological method of analysis was used to examine the core of the disaster-related experiences. Phenomenological psychology was used because this variant of phenomenology focuses on human behaviors as expressions of meaningful experiences that are obtained by descriptions from participants (Baker, Wuest & Stern, 1992; Valle &

Halling, 1989; Westers & Perters, 2004;). The qualitative data analysis was inspired by the grounded theory method, an interpretative variant of qualitative research grounded in phenomenology and symbolic interactionism. The analysis was used to uncover the potentially traumatizing core of the experiences and the characteristics of the event. When using the grounded theory method an assumption is that the assigned meaning to an event determines the response of a person to that event (Baker, Wuest & Stern, 1992).

The Van Kaam method was used to analyze the information (Baker, Wuest & Stern, 1992; Crotty, 1996; Valle & Halling, 1989; Wester & Perters, 2004). This method consists of six steps: (1) descriptive expressions are listed, and then classified into categories and ranked by frequency of occurrence; (2) these expressions are reduced to terms that describe the experience more precisely. Then, the focus shifts to the differences and similarities between experiences, in order to uncover the characteristics that are constant and the characteristics that are typical for certain subgroups (Wester & Perters, 2004); (3) the irrelevant elements (not inherent to the experience) of Step 2 are eliminated; (4) a hypothetical identification of the lived experience arises; (5) this identification is applied to some participants to test whether the description fits the data; and (6) when needed, the description is expanded or reduced. This approach resulted in a prototypical identification of the experience and a theoretical description of the traumatic core of the Ghislenghien event. The qualitative data analysis used the MAXQDA 2 software (manufactured by VERBI GmbH, Berlin, Germany) which supports researchers performing qualitative data or content analysis by helping systematically evaluate and interpret textual data.

Results

Characteristics of the Disaster Experience

There were 20 characteristics identified for the firefighters and 18 characteristics for the emergency medical services personnel. They are regrouped in respectively 14 and 13 categories and ordered according to the frequency of occurrence. It was assumed that the more frequently the characteristics were mentioned, the more prototypical they were.

Firefighters – For firefighters, experiencing the disaster in which colleagues died, made the experience of the event mainly characterized by the death of friends (Category 1: 78.7%) and to a lesser degree, by the death of other victims on the roadside and on site (Category 2: 25.5%). The event and the casualties among colleagues had made an extensive impact, and members of the involved, local fire brigade described their group as "one big family". However, after these fire and rescue interventions, also positive aspects (Category 3: 20.2%) were expressed. Firefighters mentioned that they were thankful to be alive, appreciating their job and feeling "lucky" to be still alive. They also reported that they loved their most significant others even more after response than before. Moreover, some were thankful for the psychological support (Category 9: 7.4%) that they and/or colleagues received after the responses. Some noted the impact (Category 4: 19.2%) of the intervention, during the rescue operations and afterwards – even after several months. One firefighter said, “At the beginning of the month October {nearly 3 months after the event}, I got a breakdown, it was difficult.” Others mentioned the impact on a physical level during the intervention, e.g., “I collapsed like a piece of doll.” Overall, an apocalyptic feeling and the impact on the human and physical levels sometimes resulted in an inability to speak (Category 8: 8.5%). Others were unable to talk or did not want to talk about the disaster, from the moment of the explosion several months later. One firefighter wrote: “What I saw? One page is not enough.” For some firefighters, the disaster remains their worst memory and some were not able to hold back their tears (Category 13: 3.2%) after the intervention. Sometimes, some aspects of dissatisfaction (Category 10: 6.7%) were expressed, i.e., about the disorganization on the spot.

During the rescue interventions, firefighters searched and found victims, with burns (Category 5: 18.1%), wounds (Category 6: 17.0%) and/or suffering (Category 11 : 5.3%), and screaming and/or crying (Category 12: 4.3%) for help or because of fear. Especially when driving along the way to the epicenter, they encountered people with burns lying by the roadside. Some firefighters mentioned the massive destruction (e.g., “There was nothing”) and the massive amount of victims (Category 7: 16.0%). One wrote: “As the devastation of the world after a war”. Before and at the beginning of the intervention, the firefighters did not know much about the origin of the explosion and the fire. Some of them initially heard or thought that a plane had crashed. Due to a lack of information, there was much uncertainty during the first moments after the explosion. Sometimes, the unfamiliarity (Category 6: 17.0%) with disasters in general seemed to contribute to a feeling of not knowing what to do (powerlessness), but this feeling disappeared soon after instructions were provided. Stress (Category 13: 3.2%) was evoked in a few firefighters when they were at a distance from the epicenter, and did not know much or anything about the event (in the beginning). On the spot, firefighters sometimes were at risk during the intervention. A few experienced an erroneous perception of time (Category 13: 3.2%); e.g., they did not seem conscious that time passed during the intervention. Furthermore, during the intervention, firefighters mentioned a feeling of disconnection from the reality, i.e., a feeling of not belonging to their own body. After the intervention, this feeling was experienced as not having had many emotions during the rescue operations, and wondering why not. Moreover, the feeling of working as on automatic pilot (category 14: 2.3%) also was present during the intervention.

Emergency medical personnel – The most frequent experience during emergency medical interventions related to the injured victims (Category 1: 81.3%): “The wounded persons arrived, it was terrible.” Most emergency medical personnel waited in the hospitals for the victims to arrive, saw the arrival of the first wounded and burned people (Category 6: 29.3%), who were screaming (Category 7: 24.0%)

for help, asking for "painkillers" and crying (Category 10: 13.3%) because of the immense pain. Especially the (Category 3: 33.3%) arrival of a large number of victims and the suffering was hard to witness, i.e.; "I realized the horror because the first wounded persons arrived". Many described the "overwhelming impact of the scene of a war" (Category 2: 50.7%), with the devastation of the environment and buildings, while others related the impact to the number of wounded and burned (Category 3: 33.3%). This explosion caused many casualties and some victims who still were alive asked for euthanasia, which severely impacted (Category 2: 50.7%) on the medical personnel who were trying to assist the victims. Emergency medical services and hospital staff personnel described the scene as "very shocking", e.g., "It was a nightmare", "Hopefully, I will not experience such an event again". Emergency medical personnel experienced the situation at the emergency department as "apocalyptic" and reported that they could not speak shortly after receipt of the victims, were not able to answer any questions, nor did not want to.

The experience also impacted on their personal lives, as some firefighters died (Category 4: 33.3%). Wounded and burned victims, crying and screaming (Category 10: 13.3%), are very hard to experience, e.g., "The most difficult [most painful] was at the emergency room and at the surgery room." In addition, some emergency medical personnel mentioned that they never will forget the images of what they saw. Expressions of this impact included, "I shall live with these memories my whole life" and "How to forget? Time goes by, but the memories are still very vivid". The experiences of fatalities among the firefighters also were difficult. Some of the emergency medical personnel knew the deceased firefighters on a personal level, and they felt "related" because they participated in the same rescue operation. Coping with the family members and friends (Category 5: 32.0%) of the survivors also was difficult. In addition, a few emergency medical personnel reported that they still could not believe that something like this had happened or that they simply could not accept that some of the firefighters had died.

At the beginning of the response, there was not much information available. For example, it was not known (Category 3: 33.3%) what had happened, where it had occurred and what needed to be done. The unknown evoked additional feelings of stress (Category 9: 14.7%), e.g., when preparing to depart or waiting for the arrival of the victims. Stress was triggered in emergency medical personnel and the victims because of the uncertainty about the situation of family and/or friends. Family or friends of the victims had to be dealt with, taken care of, and informed about the situation of loved ones. The likelihood of friends or family member involved, and the unknown consequences also triggered a state of restlessness (Category 11: 8.0%).

Some emergency medical personnel also noted positive aspects (Category 8: 16.0%) in relation to the disaster, and reported positively about the mutual assistance provided by emergency service personnel and citizens, and the benefits and the effectiveness of their support. They also described positive aspects of their life since the event. These aspects can be a result of experiencing the disaster. For example, "This event made it possible for me to appreciate life, to be lucky to live along and without suffering, or not that much compared to others. This event changed my perspective in a positive way." It seems that, after having been confronted with the disaster, they tried to moderate daily happenings, to be positive, and as time passes, they will learn to live with the experience. However, others also spoke about their dissatisfaction (Category 8: 16.0%) concerning various aspects related to their rescue work or related to the period after the disaster. For example, the unhealthy curiosity of bystanders during the rescue work, the media hype, the lack of psychological debriefing after the rescue operations, and the limited care and support provided for families and personnel.

During the disaster intervention, a few emergency medical personnel described that they felt like they had worked on automatic pilot (Category 12: 6.7%). Those who described this sensation felt no emotions and worked professionally as living robots. Furthermore, while working at the scene of the event, a feeling of life threat sometimes was experienced. In addition, a feeling of life threat sometimes was present when providing care of the victims and/or family. Only very few emergency medical personnel mentioned that a life threat (Category 13: 5.3%) was a concern; this can be explained by the fact that most of the respondents had been working within the hospital.

Feelings and Emotions

Firefighters – The firefighters' most frequently reported feeling was powerlessness (Category 1: 90.4%). They felt powerless because they were on the spot, and did not know whom to help first and where to start. Firefighters felt powerless when seeing all the victims; the firefighters in the first rescue squads to arrive at the scene reported consternation – the scene seemed unreal. The experience also evoked a feeling of horror (Category 2: 75.5%), and was described by a few firefighters as "an apocalypse" (Category 4: 12.8%), e.g., "I got a feeling of apocalypse when seeing the scattered bodies." Some firefighters also experienced fear (Category 3: 31.9%), especially due to the awareness that friends or family members possibly were involved. On the scene of the event, there was fear of getting involved in an accident because of the risks and of not seeing colleagues anymore. When surviving the risks, some firefighters experienced a feeling of grief (Category 5: 4.3%), due to the loss of friend. A few also reported feelings of panic, sadness, frustration resulting from the feeling of uselessness, and sympathy for the families and friends who had lost loved ones (Category 6: 1.0%).

Emergency medical services – Among emergency medical services personnel, the most frequently experienced emotion was horror (Category 1: 55.5%). A feeling of horror was experienced in relation to the arrival of large numbers of wounded and burned victims who were screaming and crying. The horrible pains and the suffering of the victims were apparent on their faces. This arrival of suffering victims also triggered a feeling of powerlessness (Category 2: 50.7%), not knowing where to start, what to do, and a feeling of being overwhelmed. Moreover, a feeling of fear (Category 3: 34.7%) also was evoked. Emergency medical personnel became frightened when they saw the flames, heard the loud sound of the explosion, thought about the emergency personnel on the scene, and saw the frightened patients. Other emotions were reported to a lesser extent. A feeling of sadness or grief (Category 4: 8.0%) was related to confrontation with the families and friends of the victims. Emergency medical personnel rarely experienced feelings of anger (Category 5: 5.3%). A few became angry because they could not understand why the event happened and why people died. When they realized that their own family members or friends were or could have been on the disaster scene, they felt panic, or they remained in despair (Category 6: 4.0%) for a long time.

Most Disturbing Aspects

Firefighters – The most shocking aspects reported by the firefighters were the casualties of colleagues and the contact with death (Category 1: 40.4%) during the intervention. They also mentioned the impact (Category 2: 29.8%) of the disaster (e.g., the amplitude of the explosion) and the number of victims and amount of damage (Category 4: 18.1%) as most disturbing aspects. Others noted that the most shocking encounters were with burned victims (Category 3: 25.3%), and the scene when they arrived (Category 5: 17.0%) e.g., "Seeing the horror which leaded to nothing." Still, others reported that the most shocking the experience was a feeling of being powerless (Category 6: 13.8%), or the confrontation with the victims (Category 7: 11.7%) they were searching for. Finally, the unknown aspects before and/or during the response (Category 8: 7.5%), and seeing the carbonized bodies and cars (Category 8: 7.5%) were shocking.

Emergency medical services – The most disturbing aspects reported by the emergency medical personnel were both the casualties among the fire fighters (Category 1: 33.3%) and the confrontation with death (Category 2: 33.3%), i.e., experiencing a contact with death or by hearing some victims asking for euthanasia. For nearly one-third of the medical services personnel, the dying of the five firefighters was especially disturbing because they knew some of them personally (Category 3: 32.0%), e.g., "Most shocking was the death of a firefighter who I knew very well". Seeing the victims in pain with burns (Category 4: 30.7%), suffering and screaming (Category 6: 18.7%), or the arrival (Category 5: 22.7%) of burned victims also was disturbing. Some others reported about the massiveness of the disaster situation in the hospital (Category 5: 22.7%). Besides what was seen or heard, the smells (Category 8: 16.0%) were disturbing. The emergency rescue personnel reported the odor of the materials (in the hospital or the mobile emergency post) and the odor proceed by the wounded victims, i.e., of burned flesh. Lastly, not knowing (Category 8: 16.0%) what to expect at the beginning of the response, and family members and friends waiting for information because nothing was known, was viewed as disturbing afterwards.

Disaster-Related Experiences after 14 months

At T2, both firefighters and emergency medical personnel reported which aspects of the disaster experience still were very present in their minds. In order to describe these aspects in a coherent narrative structure, the categories are not mentioned in the order of decreasing frequency of appearance.

Fire fighters – The most frequently described aspect in their experience was the impact both on a psychological and personal level (Category 1), e.g., "The event changed a lot of things in my family life." Some firefighters noted that they did not answer the first questionnaire because they did not feel ready (Category 5). Furthermore, the disaster had such an impact that it stayed vividly in the mind of some firefighters, e.g., "The entire intervention from the emergency call to the arrival to the end of the intervention to the return to the fire department stays very clear and very precise in my memory." Others called the response the "worst intervention ever". Still others felt turned upside down by the experience. Especially, the memories of the view at the scene (Category 2) still were difficult, e.g., when they arrived and experienced an apocalyptic feeling; or on seeing the enormous flames. The memories of the wounded or dead victims (Category 5), scattered everywhere along the road, the confrontation with death (Category 5) and the involvement of friends/family members (Category 5) as well, remained vividly in their mind. In addition, they recalled the victims who had suffered (Category 6), and the firefighters and their families who had suffered after losing some friends and/or family members including the personal impact. All these experiences contributed to the fact that some firefighters were unable to speak during the response in this lasted for at elast one year after the event. "I was never able to talk with somebody intimately about my suffering," and "On the way back, nothing is said to each other." Moreover, powerlessness (Category 4) was experienced as a result of the number of victims and the amount of devastation. As some firefighter explained: "I did not know where to start? With whom?" However, also positive aspects (Category 3) were experienced. Some firefighters spoke about the strength that grew out of the experience: they reported a feeling of personal growth, appreciating more the simple things of life, and the friendship between colleagues became more firm following the experience. Firefighters also spoke about the support they received from family or friends.

One year later (T2), they reported that: (1) the confrontation with death (Category 1); (2) the implications on friends or family members (Category 2); and (3) the death of victims, the psychological or physical impact, and the feeling of powerlessness (category 3).

Emergency medical personnel – Emergency medical staff recalled the massiveness of the damage (Category 1) and the number of victims (Category 3) at the hospital (e.g., "Seeing the victims arriving with tens together at the emergency department"). They also recalled that "victims were lying everywhere", they screamed (Category

2) for help because of the immense pain and suffering (Category 4), and because of the huge number of victims with burn injuries (Category 5). In addition, those working at the scene recalled the amount of emergency services needed and the scattered bodies. On a sensory level, sounds (Category 6) still were remembered (e.g., "That sound stays on my mind forever."). Some emergency medical personnel remained haunted by the sound of the explosion while others remembered the sounds of the ambulances and helicopters.

The most disturbing aspects mentioned one year later (T2) were the death of firefighters and/or friends or family members (Category 1), the permanent confrontation with severely burned casualties, and eventually, their death (Category 2), and victims asking for euthanasia and/or screaming because of their pain (Category 3).

Discussion

The disaster response at Ghislenghien, as experienced by firefighters and emergency medical personnel had a huge impact on the psychological and personal life of those who responded, provided care and their family/friends. Seeing the severely wounded or dead victims, colleagues, friends, or family members was horrible. Knowing about or seeing deceased friends and/or family members during or after the disaster were considered the most shocking aspects of a disaster experience. In addition, the massive amount of victims, burned, injured, screaming, and suffering was difficult to cope with and evoked feelings of powerlessness. The first minutes of "not knowing" what the disaster was about, evoked feelings of stress and restlessness. However, the experience of the disaster also turned out to be positive for some fire or emergency medical services personnel, especially when working at the disaster scene.

Compared to the firefighters, emergency medical services personnel were directly involved with the injured victims, whereas firefighters had to cope more with the physical aspects of the event (e.g., heat, smoke, fire, smell, etc.). The unusual number of injured victims resulted in a chaotic situation, which contributed to the fact that the impact of the disaster with respect to the massive human suffering was more intense in emergency medical services personnel compared to the firefighters. In contrast to the emergency medical services personnel, firefighters mostly were present on the site of the event. Firefighters are used to working on the scene of an accident or a fire, but rarely do fire fighters die during an intervention. Furthermore, firefighters more frequently reported the impact of the first encounters with the victims on the disaster site. This disaster also was characterized as an event in which friends, family members and victims died, and working on the scene was described as "risky". Both the direct contact with death and the perceived life threat appeared to be much higher in the firefighters. Time not always was consciously perceived. In addition, the apocalyptic view at the scene and the impact seems to have provoked dissociative experiences in some fire and rescue workers. The inability to speak or not wanting to speak about their experiences was related to working on the spot as a firefighter. However, after the intervention, more firefighters reported positive aspects in comparison to emergency medical services personnel.

The firefighters involved in this disaster, experienced a more direct exposure to death and more life threat than did the emergency medical services personnel. This has been documented in the literature (Laposa & Alden, 2003) when comparing the experiences of firefighters with those of emergency medical personnel: firefighters' work-related experiences are physically more dangerous and involve greater threat to their lives. Furthermore, fear of the unknown provoked psychological responses following the performance of rescue work. Previous qualitative research on the meaning of traumatizing events, as described by nurses in an ambulance service (Jonsson & Segestin, 2003) indicated the potentially traumatizing event was an encounter with the unforeseen and meaningless. Emergency medical services personnel have to cope with the potentially traumatizing event without the possibility of being prepared. Just as in some other critical daily life events, the outcome of the Ghislenghien event was extremely negative and unpredictable: 24 victims instantly killed on-scene and >100 other severely injured (burned).

Further research is needed to investigate if the first unknown minutes of an intervention may evoke a stress response. Moreover, the potentially traumatizing character of such an experience becomes clear by the fact that seeing the casualties, colleagues and friends and/or family members evokes a feeling of powerlessness and horror. This experience can be related to the concept of 'effroi de la mort', described by Lebigot (2004). The concept of "effroi" is difficult to translate, but stands for the very first state of frozen, petrified fright after a traumatic blaze; a state which precedes the development of stress and anxiety, confronting the survivor with a unique sense of loneliness, abandonment, and exclusion from the world of the living. It leaves the survivor behind without words, a disruption between the signified (death) and the signifier (symbolic representation of the real face of death) as described by some of the firefighters. Lebigot (2004) states that the human being has no representation of "the real of death" in his/her psyche; therefore, a sudden confrontation with death leaves the survivors behind without words.

The descriptions of fire and emergency services personnel involved in the Ghislenghien disaster mentioned both cognitive and emotional experiences of a disconnect from reality. This also is in accordance with the findings of Jonsson and Segesten (2003) who described the reactions of Swedish ambulance nurses who were so focused and concentrated on helping the victim that they distanced themselves from the surroundings; they were so emotionally committed that their vision narrowed. Therefore it was expected that this phenomenon will occur on an even bigger scale when being confronted with the number of victims reported in this study. Jonsson and Segesten (2003) also confirmed that helpers can have strong feelings of engagement and empathy with victims and/or family members, and, at the same time, can feel powerlessness. Their research also showed that it seems impossible to avoid these strong feelings during the contact with victims.

This analysis of the emotions experienced uncovered the fact that working as a fire fighter evoked feelings of horror and powerlessness. These feelings were more prevalent among those fire personnel who had been directly exposed to the images at the scene. A feeling of apocalypse also was experienced because of the direct encounter with the disaster environment. A feeling of powerlessness was less manifest among emergency medical personnel, probably because of the availability of a more adequate equipment and structure to help the victims. Compared to the firefighters, they were less overwhelmed by the sudden and dangerous event. In addition, they were working in their own familiar environment, i.e., the emergency department of the hospital.

The differences in reaction to the most shocking aspects seemed to be related to one's proximity to the different disaster elements. Emergency medical personnel were shocked by the screaming, pain, and suffering of the victims, including the smell of burned flesh. In contrast, firefighters were shocked primarily by the death of their colleagues. In addition, the feeling of powerlessness, the scene and the carbonized victims and cars shocked firefighters the most.

According to Moran (1998), the experience of firefighters and emergency medical personnel which involved multiple deaths is traumatic. This view was supported by the participants' answers to open-ended questions. Moreover, the positive aspect explained by some emergency personnel about the availability of materials during the response might be related to Antonovsky's sense of coherence (Engelhard, Van den Hout & Vlaeyen, 2003). Consisting of the interrelated factors namely manageability, comprehensiveness, and meaningfulness; more specifically the manageability (the present resources meet the demands) of a given situation, seems to lead to less posttraumatic stress (APA, 1994). Comparing the experiences of firefighters and emergency medical personnel, in the Ghislenghien disaster leads to the conclusion that fire fighters experience more "post-disaster positivism" than do emergency medical personnel. Even after more than one year, firefighters expressed still more "positivism". It seems that, in this case, the more traumatic the event was experienced, the more post-disaster positivism or growth was reported.

Reporting mostly horror and feelings of being totally out of control, is in accordance with criterion A2 of PTSD (APA, 1994), and since the participants involved in this study have been confronted with death, also criterion A1 of PTSD is met, hence the potential for development of post-traumatic stress reactions and PTSD is present.

Limitations

This study has a number of limitations. There has been an important drop out between both measure moments (T1 and T2) for which no unambiguous explanation is available. The informal and anecdotal data gained during the presentation of the preliminary results in the respective fire departments, indicated that over time, the involved fire and rescue personnel wanted to stop talking or bringing back memories about the disaster. Their participation in this study may have been non-random, and the emergency responders most stricken may not have been represented – or perhaps the other way around. Given that all the firefighters but three were men, it cannot be presumed that the findings also are valid for female firefighters. Furthermore, to deepen the understanding of the traumatic core of a disaster, in-depth interviews and content analyses are necessary. The timeframe and the resources available to the researchers involved in this study did not allow one-to-one contacts with all of the stricken rescuers and caregivers to be able to uncover their experiences as they were, both on the time of the disaster and in the immediate aftermath (T1) nor after more than one year (T2). Finally, since grounded theory research does not allow generalization of the research findings (Wester & Perters,2004), actual findings only contribute to the understanding of the nature and meaning of the experience of belonging to a certain population in a certain setting.

Conclusion

This study sheds light on the characteristics of the damage as experienced by firefighters and emergency medical personnel. The results of this phenomenological analysis can help the personnel from other fire, rescue, and emergency services to recognize the normality of their reaction to the potentially traumatizing aspects of a disaster and their rescue activities. These results also can be used to prepare fire and emergency services personnel to understand the possible psychological impacts upon confrontation with potentially traumatizing events. These findings also highlighted the differences between emergency medical personnel, mainly working in an intra-hospital setting, and, the fire personnel, involved in search and rescue, fire extinguishing, and providing medical first aid. This phenomenological analysis offers an important addition to the existing quantitative trauma literature in which context-specific data are less explicit. The results of this study could not confirm whether or not firefighters who contributed to the rescue of victims described less positive reactions in their rescue experience than did those involved only in fighting the fire. However, in general, firefighters reported more positive changes after the intervention than did emergency services personnel who were more involved in person-focused incidents and confronted with extensive human suffering of both victims and their families.

The assumption that the closeness of death had the most obvious impact on a person (Hagström, 1995; Lebigot, 2002), is supported. The demise of friends and/or family members was the most shocking, potentially traumatizing, aspect in this disaster experience. The unspeakable experiences of fire fighters, confronted with their own death through the direct life threat and the death of their colleagues, can be related to the concept of or a close encounter with death, leaving the survivor without words.

The result that emergency medical personnel seemed to have experienced more horror than did the firefighters could be explained by the fact that it is a more diverse and less specifically trained group, e.g., all the staff present in the hospital during the disaster had to participate in the first response at the emergency department, even staff who had never seen severely wounded people due to the fact that they were used to work in administration, kitchen, cleaning service, etc.

Authors' contributions

EDS and MK participated equally as the core research team in the conception and the design of the study, interpreted the results using the MaxQda software for qualitative analysis and drafted the preliminary manuscript of this article. EZ, RK, OVDH and JM critically reviewed the document for intellectual content and style, and have given final approval for this version to be published.

Acknowledgements

The authors thank Frédéric Daubechies, Director of the Service d'Appui Psychologique aux Intervenants for both his practical and moral support and assistance with the organization of this study. Without his help in contacting the involved fire, rescue, and emergency medical services, this research never would have been possible. Furthermore, we also would like to express our gratitude to the leaders of the fire and emergency departments and to the Union Royale des Sapeurs-Pompiers du Hainaut (Royal Union of Fire Fighters of the Province of Hainaut) more specifically for the immediate support and enthusiasm with respect to this study.

Last but not least, the authors also assure their respect and devotion to the community of fire, rescue, and emergency medical personnel, for the work they continue to do and the services they continue to deliver in these sometimes very dramatic contexts, thereby often risking their own lives.

References

American Psychiatric Association (1994). *Diagnostic and Statistical Manual of Mental Disorders (4th Ed)*. Washington, DC: APA, 1994.

Baker, C., Wuest, J., & Stern, P. (1992). Method slurring: The grounded theory/phenomenology example. *Journal of Advanced Nursing,17,*1355–1360.

Beaton, R., & Murphy, S. (1995). Secondary Traumatic Stress in Crisis Workers: Research Implications. In: Figley C (ed.), *Compassion Fatigue.* New York: Brunner-Mazel, pp 51–81.

Bravo, M., Rubio-Stipec, M., Canino, G., Woodbury, M., Ribera, J. (1990). The psychological sequelae of disaster stress prospectively and retrospectively evaluated. *American Journal of Community Psychology,18,*661–680.

Clohessy, E., & Ehlers, A. (1999). PTSD symptoms, responses to intrusive memories and coping in ambulance service workers. *British Journal of Clinical Psychology,38,*251–265.

Crotty, M. (1996). *Phenomenology and Nursing Research.* Melbourne: Churchill Livingstone.

De Soir, E. (1999). Psychologische debriefing bij brandweermannen en ambulanciers. *NVBB Magazine,146*,27–36.

Dirkzwager, A., Grievink, L., Van der Velden, P., & Yzermans, C. (2006). Risk factors for psychological and physical health problems after a man-made disaster. *British Journal of Psychiatry,189*,144–149.

Durham, T., McCammon, S., & Allison, E. (1995). The psychological impact of disaster on rescue personnel. *Annals of Emergency Medicine,14*,664–668.

Dyregrov, A. (1992). *Katastrofpsykologi*. Lund: Studentlitteratur.

Dyregrov, A., & Mitchell, J. (1993). Traumatic Stress in Disaster Workers and Emergency Personnel. In: Wilson JP, Raphael B (eds), *International Handbook of Traumatic Stress Syndromes*. New York: Plenum, 1993, pp 905–914.

Engelhard, I., Van den Hout, M., & Vlaeyen, J. (2003). The sense of coherence in early pregnancy and crisis support and posttraumatic stress after pregnancy loss: A prospective study. *Behavioral Medicine,29*,80–84.

Fullerton, C., & Ursano, R. (1997). The Other Side of Chaos: Understanding the Patterns of Posttraumatic Responses. In: Fullerton CS, Ursano RJ (Eds.), *Posttraumatic Stress Disorder. Acute and Long-Term Responses to Trauma and Disaster*. Washington, DC: American Psychiatric Press, pp. 3–15.

Green, B., Grace, M., & Gleser, G. (1996). Identifying survivors at risk: Long-term impairment following the Beverly Hill Supper Club Fire. *Journal of Consulting and Clinical Psychology,53*,672–678.

Grevin, F. (1996). Posttraumatic stress disorder, ego defense mechanisms, and empathy among urban paramedics. *Psychological Reports, 79*,483-495.

Hagström, R. (1995). The acute psychological impact on survivors following a train accident. *Journal of Traumatic Stress,8*,391-402.

Hartsough, D. (1985). Emergency organizational role. In: *Role Stressors and Supports for Emergency Workers*. NIMH; DHHS Publication N°. (ADM) 85–1408.

Herman, J. (2002). *Trauma en herstel. De Gevolgen van Geweld van Mishandeling thuis tot politiek geweld*. Amsterdam: Wereldbibliotheek, 2002.

Hodgkinson, P., & Stewart, M. (1991). *Coping with Catastrophe: A Handbook of Disaster Management*. London: Routledge.

Jeavons, S., Greenwood, K., & Horne, D. (2000). Accident cognitions and subsequent psychological trauma. *Journal of Traumatic Stress,13*,359–365.

Jonsson, A., & Segestin, K. (2003). The meaning of traumatic events as described by nurses in ambulance service. *Accident and Emergency Nursing,11*,142–152.

Kessler, R., Sonnega, A., Bromet, E., & Hughes, M. (1995). Posttraumatic stress disorder in the national comorbidity survey. *Archives of General Psychiatry,52*,1048–1060.

Laposa, J., & Alden, L. (2003). Posttraumatic stress disorder in the emergency room: Exploration of a cognitive model. *Behaviour Research and Therapy,41,*49–65.

Lebigot, F. (2002). L´effroi du traumatisme psychique. Le regarder en face ou s´en protéger. *Revue Francophone du Stress et du Trauma,2,*139–146.

Lebigot, F. (2004). Le traumatisme psychique. *Revue Francophone du Stress et du Trauma,4,*5–11.

Leon, G. (2004). Overview of the psychosocial impact of disasters. *Prehospital and Disaster Medicine,19,*4–9.

Martelli, T., Waters, L., & Martelli, J. (1989). The police stress survey: Reliability and relation to job satisfaction and organizational commitment. *Psychological Reports,64,*267–273.

Mitchell, J., & Dyregov, A. (1993). Traumatic stress in disaster workers and emergency personnel. In: J. Wilson & B. Raphael (Eds.), *International Handbook of Traumatic Stress Syndromes.* New York: Plenum,1993, pp 905–914.

Moran, C. (1998). Individual differences and debriefing effectiveness. *The Australian Journal of Disaster and Trauma Studies,1,*1-14.

Morren, M., Yzermans, J., Van Nispen, R., & Wevers, S. (2005). The health of volunteer fire fighters three years after a technological disaster. *Journal of Occupational Health,47,*523–532.

Myles G, Levine K, Ramsden F, Swanson (1990). The impact of providing help: Emergency workers and cardiopulmonary resuscitation attempts. *Journal of Traumatic Stress,3,*305–313.

Shakespeare-Finch, J., Smith, S., Gow, K., Embelton, G., & Baird, L. (2003). The prevalence of posttraumatic growth in emergency ambulance personnel. *Traumatology,9(1),*58–70.

Soeterman, R., Yzermans, C., Kerssens, J., Dirkzwager, A., Donker, G., Van den Bosch, W., & Van der Zee, J. (2006). The course of post-disaster health problems of victims with pre-disaster psychological problems as presented in general practice. *Family Practice,23(3),*378–384.

Smith, A., & Roberts, K. (2003). Interventions for post-traumatic stress disorder and psychological distress in emergency ambulance personnel: A review of the literature. *Journal of Emergency Medicine,20,*75–78.

Tugade, M., Frederickson, B., & Barrett, L. (2004). Psychological resilience and positive emotional granularity: Examining the benefits of positive emotions on coping and health. *Journal of Personality,72,*1161–1190.

Ursano, R., Fullerton, C., Vance, K., & Kao, T. (1999). Posttraumatic stress disorder and identification in disaster workers. *American Journal of Psychiatry,156,*353–359.

Valle RS, Halling S: *Existential-Phenomenological Perspectives in Psychology. Exploring the Breadth of Human Experience*. London/New York: Plenum Press, 1989.

Van Kaam, A. (1959). Phenomenal analysis exemplified by a study of the experience of really feeling understood. *Journal of Individual Psychology,15*,66–72.

Van Kaam, A. (1966). *Existential Foundations of Psychology*. Pittsburg: Duquesne University Press.

Van Kaam, A. (1969). *Existential Foundations of Psychology*. New York: Doubleday.

Weiss, D., Marmar, C., & Metzler, T., & Ronfeldt, H. (1995). Predicting symptomatic distress in emergency services personnel. *Journal of Consulting and Clinical Psychology,63*,361–368

Wagner, D., Heinrichs, M., & Ehlert, U. (1998). Prevalence of symptoms of posttraumatic stress disorder in German professional fire fighters. *American Journal of Psychiatry,155*,1727–1732.

Wester F, Perters V: *Kwalitatieve Analyse*. Bussum: Coutinho B.V., 2004.

PART 2

The Ghislenghien Gas Explosion: Empirical Research

CHAPTER 5

A Longitudinal Study on the Ghislenghien Disaster in Belgium: Strengths and Weaknesses of the Study Design and Influence on Response Rate

A. Versporten, E. De Soir, E. Zech & H. Van Oyen

Paper published in *Archives of Public Health* 2009;67:116-127.

A longitudinal study on the Ghislenghien disaster in Belgium: strengths and weaknesses of the study design and influence on response rate

Ann Versporten, Erik De Soir, Emmanuelle Zech,
Herman Van Oyen

Summary

A longitudinal study was conducted in order to assess the impact of the Ghislenghien disaster (Belgium) on physical, mental and social health, and to evaluate the prevalence of Post-Traumatic Stress Disorder (PTSD) in the affected population.

The objectives of this study were: 1) to describe the set up of the study, to report on the strengths and weaknesses of the methodology employed and its influence on response rate; and, 2) to clarify the importance of the study for the management of disasters.

The study included adults (≥ 15 years) and children (8-14 years) at risk of developing adverse health effects related to the disaster. Subjects were connected to the disaster through their geographical or professional proximity as well as connections through relatives. Questionnaires were sent by regular mail 5 months and 14 months after the disaster. Pearson Chi square tests were used to investigate whether the response rate at 14 months depended on the exposure classification.

The response rate at household level was respectively 18% (n=607 families) and 56% (n=338 families) 5 months and 14 months after the disaster. Response rate at the follow up period did not significantly differ by exposure classification.

This paper discusses the difficulties and challenges encountered during the design of the study. It discusses the determinants of response in relation to disaster related characteristics. It further provides an overview of lessons learnt and the significance of the study for the management of large scale emergencies.

Keywords: technological disaster, exposure classification, gas explosion, longitudinal study, response rate

Background

On July 30th, 2004, an accidental gas leakage occurred in a natural gas pipe under high pressure (80 bars) which passed under the industrial zone of Ghislenghien in the Walloon region of Belgium. This technological disaster killed 24 people; mainly factory workers, firefighters, and a police officer. About 132 people were wounded, many suffered severe burn injuries. Most of the wounded victims were factory workers, firefighters, police officers, and car drivers passing by at that moment on a highway 500 metres from the epicentre of the gas explosion. There was no evidence of additional chemical, air and water contamination. This disaster disrupted the life of those victims and their families physically, emotionally, and mentally (De Soir et al., 2008). The disaster can be considered to be one of the deadliest collective emergencies in Belgium since the mine disaster of the 'Bois du Cazier' in 1956 and the Innovation fire in Brussels 1967.

Introduction

Disasters cause physical harm and can result in deaths, injuries, and illnesses. Furthermore they can have a profound psychosocial impact. Nowadays, many industrialised countries have developed procedures to evaluate the consequences of disasters on the physical, emotional, and mental health of affected people. The problems experienced by disaster survivors are now interpreted in their medical and psychological context, serving as a basis for systematic and structured action by governments. In Belgium, the disaster of Ghislenghien in part has led to the scientific measurement and description of disaster-related health effects, and the improvement of post-impact health care delivery (Binder & Sanderson, 1987).

According to Noji (1995) disaster epidemiology aims to measure and describe the health effects of disasters and the factors contributing to these effects. Based on this description, this longitudinal study on the disaster of Ghislenghien evaluated the overall impact of the disaster on the general health of the affected population, with a special focus on mental health. It further provided the Belgian Federal Public Service Health, Food Chain Safety and Environment with the necessary information to improve and harmonise disaster aid structures in Belgium. The current study was carried out by the Institute of Public Health Unit of Epidemiology in collaboration with the Faculty of Clinical Psychology of the Catholic University of Louvain and the Stress and Trauma Research Centre of the Royal Military Academy.

This paper aims to report on the methodology employed, to describe the composition of the study population, the design of the questionnaires, the classification of the degree of exposure to the disaster and the health endpoints investigated. Results on the response rate are reported and compared to similar research. Strengths and weaknesses of the employed methodology are discussed. Finally, the paper provides an overview of some lessons learnt and the significance of the study for the management of large scale emergencies.

Methodology

Study population

Since victims are difficult to define and the denominator usually is unclear in the direct aftermath of disasters (Grievink, Van der Velden, Yzermans, Roorda & Stellato,

2006), our target group composed of all potential victims of the Ghislenghien gas explosion. They included residents living near the industrial site and employees of companies located on the industrial site, as well as their family members living at the same address, including children from 8 to 14 years old. We also targeted family members of deceased and injured persons. All of them were connected with the disaster through a geographical or professional proximity as well as connections through relatives.

The study population was composed using three types of records. The first were the inhabitants living up to maximum 5 km away from the explosion epicentre (n= 6382 adults and 721 children). The information was made available by the municipalities of the declared disaster area. Residents originated from the municipalities of Ath, Silly and Lessines. Secondly, 11 companies located on the industrial ground of Ghislenghien were contacted. Four companies sent administrative lists. All employees, whether or not present during the explosion, as well as students working temporarily during that period were included in the study population (n=606 adults and 96 children). Finally, the identification of victims and family members was done using a listing of people who identified themselves to an emergency service or another aid service (n=159 adults and 3 children). Those lists were made available by services concerned under the coordination of the Federal Commission of Psychosocial Help.

Household composition was not known for all families; a standard number of 6 questionnaires were sent to those households (n=158 families). Excluded from the study population were children below the age of 8 years and people who were on the site as either firemen, policemen, or from the medical intervention team. Participation to the study was voluntary. The individuals agreed to participate in the study by sending a consent form to the Institute of Public Health or by simply filling in and sending back the questionnaire.

Longitudinal study design

Auto-questionnaires were sent by regular mail services to 3,448 households in December 2004, 5 months after the disaster, The families comprised of 7,148 persons aged 15 years and older and of 820 children; aged 8 to 14 years (Table 1).

A follow-up assessment during September 2005 enabled the evaluation of changes in health impact between the period 5 to 14 months after the disaster. The second questionnaire was sent to all people who returned the first questionnaire as well as all their household members (whether or not they had replied to the 1st questionnaire). As such, we enlarged our initial sample of respondents by 238 adults and 12 children who did not respond to the 1st questionnaire. A reminder was sent in January 2006 with the aim of achieving a response rate of more than 50%.

Table 1. Response rate 5 months and 14 months after the disaster

	Questionnaires sent		Response rate	
	5 months after disaster	14 months after disaster	5 months after disaster	14 months after disaster
N households	3448	609	607 (18%)	338 (56%)
N adults (≥ 15 year)	7148	1265	1027 (14%)	579 (46%)
N children (8-14 year)	820	140	128 (16%)	74 (53%)

Table 2. Set of questions used during first and second questionnaire

	Adults ≥ 15 years		Children 8-14 years	
	5 months after disaster	14 months after disaster	5 months after disaster	14 months after disaster
Socio-demographic information*	X	X	X	X
Absenteeism at work and school	X	X	X/proxy	X/proxy
Type of exposure to the disaster	X	-	X	-
Emotional reactions during and just after the disaster (including peri-traumatic dissociation)	X	-	X	-
PTSD evaluation (23)	X	X	X	X
Post traumatic growth (24)	-	X	-	-
Description of social consequences	X	X	-	-
Description of psychosocial support/interventions by professionals	X	X	proxy	proxy
Evaluation of contacts with structures of psychosocial support	X	X	-	-
Evaluation of general health condition*	X	X	X	X
Mental health consequences (SCL90-R)* (25)	X	X	-	-
Effects on lifestyle (smoking/ alcohol use)*	X	X	-	-
Effects on drug use*	X	X	proxy	proxy
Effects on health services utilisation*	X	X	proxy	proxy

* Questions similar as those used by the Belgian Health Interview Survey (7)
* Proxy: questions filled in by a parent or a close adult

Health outcomes

Data on health outcomes were collected through two different auto-questionnaires (Table 2). One questionnaire was addressed to people aged 15 years and older. An adapted questionnaire addressed children aged 8 to 14 years, measuring the same indicators as the adults' questionnaire. Several modules of validated questions were used. As such, the same instruments on physical and mental health as those used in the 2001 Belgian National Health Interview Survey (Demarest et al., 2002) were used, with the aim of comparing the results of the present study to the data of either the total Belgian population or the population of the province of Hainaut.

With regard to mental health, somatisation, anxiety, depression, and sleeping disorders as well as Post-Traumatic Stress Disorder (PTSD) were investigated. With regard to physical health, the study addressed the use of public health services, changes in life style factors, such as smoking and alcohol use, the consumption of psycho-active drugs, and absenteeism at work or school. Some health outcomes were added during the second phase, namely post-traumatic growth which represents the positive changes in the aftermath of a crisis whereby a cognitive process is initiated to cope with the traumatic event (Hoyle & Leigh, 2000) and sense of coherence, which is the ability to comprehend, to manage and to give a meaning to the crisis (Antonovsky, 1993).

Exposure classification of the victims

Disaster victims have often been classified by the extent to which they suffered personal injury and sickness, bereavement or property loss (Dorn, 2007). Through the questionnaire, people described their degree of exposure to the disaster. The study population has been classified into 9 categories, based on the individual's proximity to the disaster. This information was only available for adults aged 15 years and older.

The 9 categories were:

- Persons injured and hospitalised ≥ 72 hours.
- Persons injured and hospitalised < 72 hours.
- Persons injured but not hospitalised.
- Persons not injured, but direct witnesses of the explosion as they saw injured or deceased victims. They were present at or nearby the site at the moment of the explosion.
- Local residents living in the surrounding communities. They had heard, seen, smelled, felt, and/or experienced the disaster from a distance.
- Family members or colleagues of deceased or wounded persons.
- Family members or colleagues of persons who could have been deceased or wounded (for example they were coincidentally not present at that moment).

- Other persons who could have been on the industrial site, but were not at that moment.

Some people were not classified: they were mainly residents exposed to the disaster through the media or they had not been exposed to the disaster in any direct or indirect way (they are not "victims"- they were absent at the moment of the disaster, e.g. on holiday).

For reasons of subsequent analyses, the 9 categories were reduced to 3 major exposure groups, following the criteria of proximity to the disaster:

- Direct witnesses who had seen human damage (SHD). They were direct 'active witnesses' of the explosion, they were present at the epicentre of the disaster. They are more prone to adverse health consequences due to the witnessing of grotesque scenes at the site and the life treat they have experienced (Norris, Friedman & Watson, 2002). They include the first four categories.
- Direct witnesses who had not seen human damage (NSHD). They were direct 'passive witnesses' of the explosion. They were not closely involved. They include category 5.
- Indirect witnesses have been indirectly exposed to the disaster through an affective proximity with a direct witness (SHD or NSHD). They include categories 6 to 9.

Results

Overall response rate

Five months after the disaster, response rate at household level was 18% (n=607 families) (Table 1). None of the family members of the 24 deceased persons participated in the study. After three months, trained psychologists contacted 43 (83%) out of 52 direct victims in order to encourage them to participate. 23 of them agreed to fill in the questionnaire, 9 declined, 11 persons could not be contacted. The response rate at household level was 17% for the residents living within 5 km of the epicentre, 23% for the employees of companies and 25% for the people who identified themselves to an emergency service or another aid service.

Fourteen months after the disaster, the response rate at household level was 56% (338 families); this is 9.8% of the initially study population who received a questionnaire five months after the disaster. We captured 32 adults and 3 children who responded to the 2nd questionnaire only.

Distribution rate by exposure classification

At 5 and 14 months respectively, the overall distribution by exposure classification was 8.2% (n=84) and 8.1% (n=47) for direct witnesses SHD. Among them, 38% (n=31) were injured and 63% were not (n=53) at 5 months (Table 3). At 5 and 14 months respectively, the overall distribution by exposure classification was 58.1% (n=597) and

55.3% (n=320) for direct witnesses NSHD and 33.7% (n=346) and 31.1% (n=180) for indirect witnesses. We investigated whether the response rate at 14 months depended on the exposure classification. The comparison is of interest in drawing inference on further research results from the data. The response rate did not differ significantly by exposure classification (Pearson Chi square test: NS at p=0.47).

Table 3. Distribution of adults (>=15 years old) by exposure classification, exposed to the disaster by geographical (resident) or professional (working in company) proximity 5 months and 14 months after the disaster

	Resident		**Working in company**		**Total**	
	5 months N (%)	**14 months N (%)**	**5 months N (%)**	**14 months N (%)**	**5 months N (%)**	**14 months N (%)**
1) Direct witness, SHD*	52 (5.8%)	27 (5.6%)	32 (21.8%)	20 (21.1%)	84 (8.2%)	47 (8.1%)
Injured & hospitalised >72h	3	2	18	11	21	13
Injured & hospitalised <72h	1	1	6	4	7	5
Injured, not hospitalized	1	0	2	2	3	2
Direct witness of human damage	47	24	6	3	53	27
2) Direct witness, NSHD*	579 (65.8%)	307 (63.4%)	18 (12.2%)	13 (13.7%)	597 (58.1%)	320 (55.3%)
3) Indirect witness	249 (28.4%)	124 (25.6%)	97 (66.0%)	56 (58.9%)	346 (33.7%)	180 (31.1%)
Close person or colleague dead or injured	44	22	64	34	108	56
Close person/colleague could have been hurt, but was not	33	15	2	2	35	17
Others, could have been on the site but were not	158	81	30	20	188	101
No classification	14	6	1	0	15	6
Others (Quest. 2)		26 (5.4%)		6 (6.3%)		32 (5.5%)
Total	880 (100%)	484 (100%)	147 (100%)	95 (100%)	1027 (100%)	579 (100%)

* SHD =Seen human damage, NSHD =Not seen human damage

Discussion

The study population included all people who could possibly have been touched by the disaster. No exclusion criteria were established. The overall response rate at household level at 5 months was 18%. In order to increase the response rate among the direct victims, they were contacted by psychologists who tried to motivate them to participate in the study. But the majority of the people who declined to participate did not want to be confronted with the disaster again. Finally, our study population represents only 8% direct witnesses, SHD (n=84). Among them, only one third was injured (38%) and 63% were not. Generally, about 132 people were known to be injured during the disaster; however the database could only capture 31 of them.

The bureaucratic and political problems that led to the delay in the onset of the investigation could account for this low number of direct witnesses SHD in the database. This pitfall raises question on the study representativeness. The possible selection bias could result in an under representation of severely affected victims. If we would have been able to capture most of the injured victims, knowing that they are more likely to participate in a study on disasters (Grievink et al., 2006), the investigators probably would have increased the response rate for this subpopulation significantly.

Fourteen months after the disaster, the response rate at household level was 56%. A comparable response rate from all different sub-groups within a sample is desirable. The response rate in this study did not differ significantly by exposure classification. This contrasts the expectation of a differential response rate among the 3 exposure groups as each of them may be using different coping mechanisms to the disaster effect and may feel differently affected by the consequences of the disaster.

The low response rate for the first questionnaire necessitates careful interpretation of subsequent results. Differences in respondent characteristics may lead to bias in prevalence estimates and bias in associations. Studies of representative samples of the population are theoretically the best approach to examine the prevalence of health effects (Lang et al., 2007). A high response rate is essential to ensure the representativeness and generalisability of the results of the study. The methods used to contact the population need to be considered. A cross-sectional study, conducted 18 months after the AZF factory explosion in Toulouse (France) among adults over 18, reached a participation rate of 60%. They used questionnaires that had to be completed during a face-to-face interview. Face-to-face interviews or telephone surveys have been reported to have higher response rates than mailed questionnaires (Lang et al., 2007). However, a response rate of 20% is considered normal for a postal questionnaire (WHO, 1996). On the other hand, it has been found that when survey methods such as questionnaires are used, then the return rates are much lower than in non-disaster settings (Logue, Melick & Hansen, 1982). Similar research on disasters in the Netherlands has reported higher response rates. Survey participation rate 3 weeks after a major firework disaster in Enschede was 26% (adults residents ≥ 18 years old) (Grievink et al., 2006). The higher response rate could be a result of the shorter time interval between the disaster and the first measurement (3 weeks instead of 5 months in the present research). Response rate at 18 months was 54% among immigrants and 81% among native Dutch persons (Dijkema, Grievink, Stellato, Roorda & Van der Velden, 2006) with an overall response rate of 72% (Van Den Berg, Van der Velden, Joris, Stellato & Grievink, 2006). They concluded that health problems were associated with higher response among non-Western immigrants and with lower response among the native Dutch.

Not responding to the present study could have 3 different causes. First, people might not have been able to answer the questionnaire due to hospitalization or recovery, be it physical or emotional. A lot of them may not have wanted to be confronted once again with the disaster. This hypothesis is more likely among direct witnesses SHD. This would lead to an underestimation of the health outcomes. Secondly, it is possible that those living further away from the epicentre of the disaster have not filled in the questionnaires because they did not feel involved or concerned with the disaster. This could apply to direct witnesses NSHD as they "could" have experienced the disaster from a distance. This applies also for indirect witnesses. It could lead to a non-response bias related to the exposure. Those who participated could be different from those who did not. If so, it could have a negative influence on the generalisability of the study results for that study group. Next, the study was not integrated in the global management of the disaster. Several institutions approached the affected people and asked for information about the disaster requiring a lot of paperwork to be filled in. People might have felt disappointed about the management of the disaster by the government. Some of the victims have been discouraged by caregivers of psychosocial support structures to fill in yet another questionnaire. At the same time, there were parallel studies, not financed by the government, which gave an uncoordinated impression to the population and might have lead to confusion. All of this might have had a negative influence on the response rate.

Researchers encountered other difficulties which could have influenced the response rate. Firstly, it took about 4 months of negotiations with the authorities before the study was approved. The time interval between the gas explosion and the first point of measurement was 5 months. Secondly, the researchers dealt with difficulties in the collection of the necessary addresses of the affected people. There was a lack of co-ordination between the different governmental departments (both local and federal), or organizations involved. Each of them claimed the exclusive right to manage the confidential data on the disaster victims. Several address lists of victims were generated in parallel, and they were never really compared by one governing body. As a consequence, the list of direct victims who were present at the site during the explosion was not made available in time; moreover, not all direct victims could be contacted. Finally, the household composition at 5 months was not known for 158 families. A set of 6 questionnaires had been sent to those households. A standard household composition counts on average 4 family members. This fact could have led to an underestimation of the true response rate due to those missing denominator data.

Several studies describing the impact of disasters on (mental) health deal with methodological problems which could be related to the use of different definitions concerning the classification of the different exposure groups (Brewin, Andrews & Valentine, 2000; Raphael & Wilson, 1993). The exposure classification of the Ghislenghien disaster victims is related to the individual's proximity to the disaster. The hypothesis was made that health problems of the defined groups would initially depend on the extent to which they witnessed or suffered personal injury, secondly on the degree of geographical distance to the explosion, and thirdly on the type of involvement through family or close connections. Disaster victims often have been classified by the extent to which they suffered personal injury, bereavement or property loss (Dohrenwend et al., 1981). Others differentiated groups according to the type of involvement, their function, their capability, their self-control and certain extraneous factors (Taylor, 1987). The classification made in our study lies closely to the classification used in the Volendam fire disaster, in the Netherlands, where the different groups depended on the degree and type of losses experienced (Dorn, 2007).

Limitations of the study in summary

The sudden aspect of this technological disaster has put time pressure on the development of the study protocol as we dealt with complicated socio-political and legal issues. Next, we faced a number of methodological problems, this includes the definition of the entire potentially 'affected' population. In addition to the direct victims being on the site of the disaster, a rather large group of residents were included into the study population in order not to miss any potentially affected victim. It was further difficult to contact and motivate the most affected victims (wounded and hospitalized participants, family members of deceased persons). None of the family members of deceased persons answered the questionnaire.

Strengths of the study

The first strength of the study is that we have included several sets of validated and widely used questions on mental, physical and social health which also have been used by the Belgian health interview survey (Demarest et al., 2002) as a reference population. Secondly, our target group composed of all potential victims including family and children. This added value, given that traumatic experiences and stressful life events frequently involve the entire support system of the victim (Dorn, 2007). Finally, people may answer in a different way in the aftermath of a disaster due to the social impact on the public as a whole inducing observation bias (attention given to the affected population, media). But, in order to avoid observation bias during the second phase, the 2nd questionnaire was sent 14 months instead of 12 months after the disaster when a lot of attention was given in the media due to the one-year memorial of the disaster.

The significance of the study for the management of large scale emergencies in Belgium

This research provides knowledge about the long-term health consequences of technological disasters. Recommendations presented to the authorities stress the need to take care of and to follow up the victims. Local general health practitioners, psychosocial actors, other professionals and the authorities should be more involved and their involvement should be structured into future disasters. This requires a multidisciplinary and proactive approach (Versporten, Zech, De Soir & Van Oyen, 2006).

As in other countries, the concept of emergency planning in Belgium is regularly evaluated. Since 1997, the Federal Public Service for Health has developed a Psycho Social Intervention Plan (PSIP) within the framework of a medical-sanitary aid service during collective emergencies. The disaster of Ghislenghien forced the Belgian authorities to re-evaluate their approach towards the management of disasters. It was recommended that a federal structure should be set up to deal with the coordination and gathering of information related to the victims of a disaster whereby the physical, mental and social health of the potentially affected population is "systematically" assessed and monitored. Our study took so long to be set up because no such structure existed. An independent structure, such as the National Institute of Public Health could be charged with the evaluation of its intervention; this institute has the experience in handling sanitary crisis situations in Belgium. A systematic evaluation strategy should exist, as in other Western countries such as the Netherlands and the United States. Their role should be to evaluate the health consequences in the aftermath of a disaster in order to evaluate the efficacy of the medico-psycho-social interventions that have taken place.

Nowadays, psychosocial aid is a formal part of the second discipline of the emergency and intervention plan as outlined in the PSIP scenario (Royal decree concerning the emergency and intervention planning, February 16 2006) (Allaert, 2008). As such, the PSIP aims to provide quality of care provided to affected people through the use of a standard reference frame at the federal level. Furthermore, it also had an effect on the tasking priorities of the existing Psycho Social Managers (PSMs). The collaboration of PSMs, whose function it is to co-ordinate the psychosocial assistance for disaster survivors would of course be most appreciated if such a systematic evaluation was (ideologically and financially) supported. Indeed, their task is on the one hand preventive, whereby the PSMs are constructing a psychosocial network on the basis of the existing psycho-social services. On the other hand, their role is operational, whereby the PSMs are coordinating the psycho-social care of disaster victims on the basis of the PSIP. In this way, the local services can be more efficiently organized, as they can use a specific coordination structure that always assesses the needs of victims, the available resources and the actions taken toward the benefit of victims. Thus, PSMs could help in dealing with the collection of important information on the victims that would be used in the identification of potentially affected victims and their inclusion in a study on their health.

Authors' contributions

EDS, AV, EZ and HVO participated equally as the core research team in the conception and the design of the study, interpreted the results and drafted the preliminary and the end reports for the Federal Public Service for Health, Food Chain Safety and Environment. HVO was central in the acquisition of funding and assured the general supervision of the core research team. EZ, EDS and AV collaborated in the collection of the data. AV statistically analyzed the data. EDS drafted the manuscript of this article and collected relevant literature references. RK, OVDH and JM critically reviewed the document for intellectual content and style, and have given final approval for this version to be published.

Acknowledgments

The study was funded by the Federal Public Service Health, Food Safety and Environment. We also thank Prof. Dr Rolf Kleber (Utrecht University) and Prof. Dr Onno van der Hart (Utrecht University) who provided advice with respect to our conceptualisation of psychological trauma and the importance of peritraumatic dissociation for inclusion in the questionnaire.

References

APA, American Psychiatric Association (1994). Diagnostic and Statistical Manual for Mental Disorders. 4th Edition ed. Washington D.C.

Allaert, E. (2008). *Psychosociaal management binnen de discipline "medische, sanitaire en psychosociale hulpverlening" in de noodplanning.* Eindverhandeling stage niveau A. FOD Volksgezondheid, Veiligheid van de Voedselketen en Leefmilieu, Directoraat-generaal Basisgezondheidszorg en Crisisbeheer

Antonovsky, A. (1993). The structure and properties of the sense of coherence scale. *Social Science and Medicine,36(6)*,725-733.

Binder, S., & Sanderson, L. (1987). The role of the epidemiologist in natural disasters. *Annals of Emergency Medicine,16(9)*,1081-4.

Brewin, C., Andrews, B. & Valentine, J. (2000). Meta-analysis of risk factors for post-traumatic stress disorder in trauma-exposed adults. *Journal of Consulting and Clinical Psychology,68(5)*,748-66

Demarest, S., Van der Heyden, J., Gisle, L., Buziarsist, J., Miermans, P., Sartor, F. et al. (2002). Belgian Health Interview Survey - Results. IPH/EPI REPORTS N°25. Brussels: Scientific Institute Public Health.

De Soir, E., Zech, E., Versporten, A., Van Oyen, H., Kleber, R., Van Der Hart, O. & Mylle, J. (2008). Prédiction de l'ESPT lors d'une catastrophe technologique. *Revue Francophone du Stress et du Trauma,8,*211-224.

Dijkema, M., Grievink, L., Stellato, R., Roorda, J., & Van der Velden, P. (2005). Determinants of response in a longitudinal health study following the firework-disaster in Enschede, The Netherlands. *European Journal of Epidemiology,20(10),*839-847.

Dohrenwend, B., Dohrenwend, B., Warheit, G., Bartlett, G., Goldsteen, R., Goldsteen, K. et al. (1981). Stress in the community: A report to the President's Commission on the Accident at Three Mile Island. *Annals of the New York Academy of Sciences,365,*159-174.

Dorn, T. (2007). *Health impact of the Volendam disaster.* Utrecht: NIVEL.

Derogatis, L. (1977). SCL-90_R (revised) version Manual I.

Grievink, L., Van der Velden, P., Yzermans, J., Roorda, J., & Stellato, R. (2006). The importance of estimating selection bias on prevalence estimates shortly after a disaster. *Annals of Epidemiology,16(10),*782-8

Health Interview Surveys (1996). Copenhagen, World Health Organisation

Hoyle Leigh (2000). Stress: Positive and Negative Effects. Posttraumatic Growth: Positive Changes in the Aftermath of Crisis. *American Journal of Psychiatry,157(10),*1712.

Lang, T., Schwoebel, V., Diene, E., Bauvin, E., Garrigue, E., Lapierre-Duval, K. et al. (2007). Assessing post-disaster consequences for health at the population level: experience from the AZF factory explosion in Toulouse. *Journal of Epidemiology and Community Health,61(2),*103-107.

Logue, J., Melick, M., & Hansen, H. (1981). Research issues and directions in the epidemiology of health effects of disasters. *Epidemiology Review,3,*140-62

Noji, E. (1995). Disaster epidemiology and disease monitoring. *Journal of Medical Systems, 19(2),*171-174.

Norris, F., Friedman, M., & Watson, P. (2002). 60,000 disaster victims speak: Part II. Summary and implications of the disaster mental health research. *Psychiatry,65(3),*240-260.

Raphael, B., & Wilson, J. (1993). Theoretical and intervention considerations in working with victims of disaster. International handbook of traumatic stress syndromes (pp. 105-117).

Taylor, A. (1987). A taxonomy of disasters and their victims. *Journal of Psychosomatic Research,31(5),*535-544.

Tedeschi, R., & Calhoun, L. (1996). The Posttraumatic Growth Inventory: measuring the positive legacy of trauma. *Journal of Traumatic Stress,9(3),*455-471.

Ursano, R., & Fullerton, C. (1997). *Trauma, time and recovery. Posttraumatic stress disorder. Acute and Long-term responses to trauma and disaster.* Washington DC: American Psychiatric Press, pp. 269-275.

Van Den Berg, B., Van der Velden, P., Joris, Y., Stellato, R., & Grievink, L. (2006). Health-related quality of life and mental health problems after a disaster: are chronically ill survivors more vulnerable to health problems? *Quality of Life Research,15(10)*,1571-1576.

Versporten A, Zech E, De Soir E, Van Oyen H. Etude de l'impact de la catastrophe de Ghislenghien sur l'état de Stress Post-Traumatique et sur la santé de la population. IPH/EPI REPORTS Nr. 2006-025. 2006. Brussels, Belgium, Scientific Institute of Public Health.

CHAPTER 6

Degree of Exposure and Peritraumatic Dissociation as Determinants of PTSD Symptoms in the Aftermath of the Ghislenghien Gas Explosion

E. De Soir, E. Zech, A. Versporten, H. Van Oyen, R. Kleber, J. Mylle & O. van der Hart

Paper published in *Archives of Public Health*, 73:21 (13 April 2015)
doi 10.1186/s13690-015-0069-9

Degree of exposure and peritraumatic dissociation as determinants of PTSD symptoms in the aftermath of the Ghislenghien gas explosion

Erik De Soir, Emmanuelle Zech, Ann Versporten, Herman Van Oyen, Rolf Kleber, Jacques Mylle, Onno van der Hart

Summary

This chapter investigates risk factors for the development of posttraumatic stress symptoms in the different survivor groups involved in a technological disaster in Ghislenghien (Belgium). A gas explosion instantly killed five firefighters, one police officer and 18 other people. Moreover, 132 people were wounded among which many suffered severe burn injuries.

In the framework of a large health survey of people potentially involved in the disaster, data were collected from 3,448 households, of which 7,148 persons aged 15 years and older, at 5 months (T1) and at 14 months (T2) after the explosion. Hierarchical regression was used to determine the significant predictors and to assess their proportion in variance accounted for.

The degree of exposure to the disaster was a predictor of the severity of posttraumatic stress symptoms. Peritraumatic dissociation appeared to be the most important predictor of the development of posttraumatic stress symptoms at T1. But at T2, posttraumatic stress symptoms at T1 had become the most important predictor. Dissatisfaction with social support was positively linked to development of posttraumatic stress symptoms at T1 and to the maintenance of these symptoms at T2. Survivors who received psychological help reported significant benefits.

In harmony with the findings from studies on technological disasters, at T1 6,0% of the respondents showed sufficient symptoms to meet all criteria for a full PTSD. At T2, 6,6% still suffered from posttraumatic stress symptoms. The symptoms of the different victim categories clearly indicated the influence of the degree of exposure on the development of posttraumatic stress symptoms. Problems inherent to retrospective scientific research after a disaster are discussed.

Keywords: technological disaster, PTSD symptoms, peritraumatic dissociation, social support, psychological help

Background

In Western countries, 5% of all deaths are due to the consequences of aggressive or unnatural, technological, events (Ursano, Caughey & Fullerton, 1994). These events may cause posttraumatic stress symptoms (Maes et al., 1998a; Maes et al., 1998b). When these symptoms lead to such a severe suffering that they hamper social, familial or professional functioning for more than one month, they are indicative of PTSD (APA, 1994). Studies indicated that the prevalence of PTSD during the first year after a disaster is 1 to 11% (Foa, Keane & Friedman, 2000; Heinrichs et al., 2005; Kessler, Sonnega, Bromet, Hughes & Nelson, 1995; Ursano, Caughey & Fullerton, 1994). Disaster victims are often difficult to define and the denominator is usually unclear in the direct aftermath of disaster. For this reason, disaster research often focuses on the residents of the official disaster area (Yzermans & Gersons, 2002) or survivors who had to be relocated after the disaster (Grievink, Van der Velden, Yzermans, Roorda & Stellato, 2006). In other cases (Dorn, Yzermans, Kerssens, Spreeuwenberg & Van der Zee, 2006), victims are identified through the medical chain and as a function of their medical condition. The posttraumatic consequences of disasters (i.e., the percentage of survivors with PTSD) depend on the type of exposure to it. For instance, two months after accidents with serious injuries 35% of survivors met PTSD criteria (Bonanno, Galea, Bucciarelli &Vlahov, 2006; Perry, Dean & Krenzelok, 1992), five to eight weeks after the disaster depending on the proximity to the World Trade Center site 7 to 20% respondents met the PTSD criteria (Galea et al., 2002), and ten years after a disaster on an oil platform 25% continued to satisfy PTSD criteria (Hull, Alexander & Klein, 2002). This indicates that there exists a relationship between the severity of exposure and the mental health condition, called dose-relationship, which could serve as a basis for victims classification. Severe exposure such as threat of life, the confrontation with injury and human losses or severe initial stress reactions may be considered as event-specific risk factors (City Health Information, 2014; Galea, Nandi & Vlahov, 2005).

In Belgium, the impact of two major accidents (Maes, 1998a; Maes, 1998b) has been investigated: (1) a sudden fire blaze in a ballroom in which about 450 guests were celebrating the 1994 New Year Eve (Switel Hotel, Antwerp, Belgium) with 15 people killed and more than 160 with burn injuries; and (2) a massive car pile-up in 1996 (Nazareth, Belgium) involving about 200 cars and trucks, in which 10 people died and 50 others were injured. The Composite International Diagnostic Interview (CIDI) was used between 7 and 9 month after these potentially traumatizing events in a study group of 185 victims; 130 fire and 55 car accident victims. The incidence of PTSD for these accidents appeared to be 26.2% for the ballroom fire blaze and 16.4% for the massive car pile-up.

The varying degrees of PTSD in disaster survivors show that scientific research is not univocal with regard to the determinants of PTSD. Many studies showed methodological shortcomings, which might be inherent to research on the impact of disasters on (mental) health: the suddenness of a disaster, the lack of control groups (Ursano & Fullerton, 1997), the different definitions of the concepts used (regarding the type of traumatizing event, psychological disturbances, dissociation, etc.) or the victim categories (Brewin, Andrews & Valentine, 2000; Hoogduin, Van Minnen, Verbraak & Van de Griendt, 2002; Miller, 2003; Tierney, 2000; Ursano & Fullerton, 1997). In the past few years, much importance has been attributed to the study of predictive factors concerning the development of psychological disturbances, such as the role of several phenomena occurring during or immediately after the potentially traumatizing impact. Reactions experienced immediately at or after the moment of the event, such as peritraumatic dissociation, extreme anxiety, panic, and negative emotions, appear to be important predictors of PTSD symptoms (Raphael & Wilson, 1993). Symptoms of peritraumatic dissociation are thought to involve cognitive alterations and alterations of perceptual functioning at the moment of, or directly after the event (Bernat, Ronfeldt, Calhoun & Arias, 1998; Marmar, Weiss, Metzler & Delucchi, 1996). A comprehensive meta-analysis (Ozer, Best, Lipsey & Weiss, 2003) revealed that peritraumatic dissociation turned out to be the best PTSD predictor in comparison with other predictors such as earlier traumatization, earlier psychological well-being, familial antecedents of psychopathology, life threat felt during the traumatic event, social support and emotional reactions occurring during or right after the event (i.e. negative peritraumatic emotions such as guilt or self-blame).

Since disasters always strike networks of people, it is also important to investigate the effect of social support. Previous research (Marmar, Weiss, Metzler & Delucchi, 1996; Ozer, Best, Lipsey & Weiss, 2003; Ursano & Fullerton, 1997) found that social support counters the development and maintenance of distress and mental disorders, in particular PTSD. Social sharing occurs in 80 to 95% of all emotional episodes and usually develops in the period immediately following an emotional event (Guay, Billette & Marchand, 2006; Rimé, Philippot, Boca & Mesquita, 1992). Modally, the sharing of an emotional episode is repetitive and addresses several successive recipients. Experimental studies confirmed emotional exposure to cause the social sharing of emotion (Finkenauer & Rimé, 1998). Emotion sharing was consistently found to hold a positive linear relationship with the intensity of the emotional experience (Finkenauer & Rimé, 1998; Luminet, Bouts, Delie, Mainstead & Rimé, 2000).

The current study aimed at 1) assessing the prevalence of posttraumatic stress symptoms in survivors of the Ghislenghien gas explosion, in their family members as well as in family members of deceased victims, 2) focusing particularly on six predictors of posttraumatic stress symptoms: the frequency and the severity of posttraumatic stress symptoms, the type of exposure, the level of peritraumatic dissociation, the received social support and the professional help.

On July 30th 2004, a leakage caused by a drilling machine in a high pressure gas pipe, which passed under the industrial zone of Ghislenghien (Belgium) created a persistent gas smell and alerted the employees of one of the factories. When the fire services arrived on-scene, an enormous explosion took place and instantly killed 24 people. Only two firefighters from the first crew survived the initial blast and 132 people were wounded. An impressive column of fire rose into the air and the heat was felt up to two kilometers away from the explosion site. Debris from the gas pipe and buildings was projected up to six kilometers away from the epicenter; up to 16 km from the explosion, air vibrations were registered. A wide area was affected by the largest technological disaster that Belgium ever knew since the mine disaster of Marcinelles (1956) in which 256 coal miners were killed by an underground fire.

Methods

Design of the study

In the framework of a health survey of people involved in the disaster, inhabitants living maximum five kilometers away from the explosion epicentre and employees of all companies located on the industrial site of Ghislenghien were contacted. The target group comprised 3,448 households, totaling 7,148 persons aged 15 years and older on August 1, 2004. All of them, as well as all their family members living at the same address and the family members of all deceased persons were invited to participate in the study. Fire, police, and emergency medical services personnel who took part in the rescue operations were excluded from the survey as they were the target of a specific investigation. Participation was voluntary. Questionnaires were sent by regular mail 5-months after the explosion (T1). Injured persons who did not react within one and a half months were contacted by telephone by experienced psychologists. Fourteen months after the disaster (T2), each person who responded to the first questionnaire received a second questionnaire for the follow-up of the impact of the disaster. To guarantee anonymity of the respondents and respect of the law on protection of private life, each questionnaire was given a unique, anonymous identification number.

Measures

Degree of exposure

Twenty-three questions (yes-no) assessed how respondents had been involved in the disaster, allowing to elaborate a classification of the degree of exposure. This classification contained 3 main categories divided into 9 subcategories. The first category comprises the primary victims; i.e. persons who had been directly exposed to the disaster and were direct witnesses of human damage. Primary victims were subdivided in 5 subcategories: 1) injured and hospitalized for more than 72 hours; 2) injured and hospitalized less than 72 hours; 3) injured but not hospitalized; 4) not injured but were direct witnesses of human damage (injured or deceased); and 5) direct witnesses not exposed to human damage, i.e. direct witnesses of the explosion effects (heat, blast, etc.). The second category encompasses the secondary, indirect victims; i.e. who had been indirectly exposed to the disaster by being related to a primary victim. It includes family members or colleagues, either 6) of people who got injured or killed by the disaster, or 7) people who could have been injured or killed by the disaster but were for some reason not present on site). The third category regroups the tertiary victims; i.e. people classified as 8) could have been exposed directly to the disaster but were not because (accidentally not present on site, or 9) not belonging to any of the previous categories

This classification constitutes an ordinal scale based on a "dose" dimension. The first category has been exposed to the highest dose of (life threatening) stress and category 9 to the least dosage. To a certain extent we can consider this scale as a continuous scale "compressed" into 9 ordered categories (Winship & Mare, 1984).

Peritraumatic dissociation

Peritraumatic dissociation was measured using the Peritraumatic Dissociative Experiences Questionnaire (PDEQ; Marmar et al., 1999). Participants rated on a 5-point Likert scale the extent to which they respectively endorsed the 10 PDEQ-statements regarding their reactions during and immediately after the event. Since internal consistency was adequate ($\alpha = .85$), an index of psychoform peritraumatic dissociation was computed by summing the item scores. The higher the total score, the higher peritraumatic dissociation.

Posttraumatic stress symptoms

The design of the study did not allow for PTSD diagnosis by clinicians. Therefore, an assessment of posttraumatic stress symptoms with the QE-PTSD, a French questionnaire, developed at the Université Catholique de Louvain and closely matching the DSM-IV criteria for diagnosing PTSD symptoms was used (Sydor,1998). Criterion A1 (objective exposure to a potentially traumatic event) was assessed by questions on the type of exposure to the disaster described above. Three dichotomous questions (yes-no) addressed Criterion A2 (subjective exposure to the potentially traumatic event: intense fear, helplessness, horror). Criterion B (intrusions) was assessed by 8 items, and Criterion C (avoidance and numbing) and D (hyperarousal) by 7 items each. All these items had to be rated by the respondent on 5-point Likert scales (0 = not at all; 4 = a great deal). The threshold for a response to be considered positive was 2. Four items addressed Criterion F (trauma-related distress and impairment in social life, work, or daily activities (yes-no). Finally, three items evaluated (1) whether symptoms had lasted for more than a month (yes-no; Criterion E), (2) whether symptoms were still apparent (yes-no), and (3) when symptoms had started (month and year). Three types of variables were calculated: (1) a dichotomic index (satisfies vs not) for each PTSD criterion (showing 1 symptom for criterion B, 3 for C, and 2 for D) plus a PTSD index (criteria A-F satisfied or not), (2) a continuous variable of PTSD severity (sum of items scores B, C, and D endorsed), and (3) continuous variables representing the number of criteria met for each cluster of symptoms (B, C, and D). This QE-PTSD instrument had been tested on various samples of respondents after they had been exposed to a potentially traumatizing event (Mutter, 1998; Sydor, 1998; Sydor & Philippot, 1996a; Sydor & Philippot, 1996b). The B, C and D criteria scales, as well as the total PTSD severity index, showed good internal consistency, respectively Cronbach's alpha = 87, .76, .86, and .93 in Mutter's sample (Mutter, 1998), and .90, .83, .91, and .91 in the present sample.

Social sharing, social support, and professional support

The three dimensions usually addressed in social sharing of a given emotional event have been used in our study: latency, extent (number of repetitions), and number and type of targets (Finkenauer & Rimé, 1998; Rimé, Philippot, Boca & Mesquita, 1992). Latency of the first sharing, extent of sharing and number of targets with whom they had shared was rated on a 7-points scale. The next question (yes/no) addresses the professionally provided support: Did you receive psychological help (personally or by telephone) after the event? Finally, a question focused on the support the involved persons offered to themselves to other victims: Did you provide support for other people involved in the disaster?

Results

Participation rates

At T1, the response rate at household level was 607 (18%) and at individual level, valid data were obtained from 1,027 adults (14%). At T2, valid data were obtained from 579 persons belonging to 338 households (46%). None of the family members of the 24 deceased persons participated in the study. Average age of the respondents was 45 years (SD = 16.6; range 15–92) and half of the sample (49.8%) were men. Nine percent of the participants (N = 95) at T1 were full time students. Among the remaining participants, 35.0% had completed higher education, 35.4% higher secondary education, and 29.5% had not finished secondary education.

PTSD symptoms

For the assessment of PTSD symptoms, only people with complete data for at least one of the criteria (A1, A2, B, C, D, E or F) were taken into account. Some 1,4% of the respondents did not meet Criterion A1 because they were not at all involved in the disaster (e.g., being on holiday) (Table 1). They were excluded from further analysis. About 91% of the respondents satisfied criterion A2. All respondents (100%) satisfied criterion B and there was no significant decrease over time since all participants were still satisfying it at T2. Criterion C was satisfied by 9% of the respondents at T1 and the prevalence was 8% at T2. Nevertheless, 51% of those who satisfied this criterion at T1 (n = 23) did not so at T2 while 4% of the respondents (n = 20) had developed this type of symptoms at T2. The proportion of recovery at T2 was thus higher than the proportion of development at T2, $\chi^2(1, N = 521) = 119.78$, $p < .0001$. It should also be noted that some items, such as the efforts to avoid thinking about the event, were quite frequently endorsed (20%). Criterion D was satisfied by one third of the respondents (34%) at T1 and the prevalence stayed similar at T2 (33%). Nevertheless, 31% of those who showed hyperarousal symptoms at T1 (n = 60) did not show them anymore at T2, while 13% (n = 44) had developed this type of symptoms at T2. The proportion of persons who recovered from hyperarousal at T2 was thus higher than the proportion of persons who developed hyperarousal at T2, $\chi^2(1, N = 527) = 169.72$, $p < .0001$. It is worth noting that all respondents reported being more on edge and more watchful since the disaster. Criterion E (onset and duration) was satisfied by 52% of the participants at T1 and 53% at T2). Here, probably due to the longer time frame, the proportion of people who stopped satisfying this criterion at T2 (25%) was smaller than the proportion of who started endorsing this criterion at T2 (33%), $\chi^2(1, N = 503) = 88.15$, $p < .0001$. Finally, criterion F was satisfied by 26% of the respondents at T1 and 18% at T2. The proportion of persons who recovered at T2 was higher (57%, n = 81) than the proportion of new onsets of dysfunction at T2 (9%, n = 33), $\chi^2(1, N = 510) = 78.74$, $p < .0001$.

Table 1 PTSD criteria satisfied after the Ghislenghien disaster (DSM IV, 1994),

	T1 : After 5 months		T2 : After 14 months		p-value ▪-test
	N = 1027	% valid	N = 579	% valid	
Criterion A1 : objective exposure	**1012**	**98.6**	**541**	**98.9**	
Criterion A2 : subjective exposure	**897**	**91.3%**	**485**	**92.2**	
Intense fear	503	56.8%			
Powerlessness	745	81.0%			
Horror	683	74.9%			
Criterion B : Intrusions	**978**	**99.9%**	**571**	**99.8%**	**ns**
Criterion C : Avoidance	**87**	**9.0%**	**48**	**8.4%**	**.0001**
Criterion D : Hyperactivation	**340**	**34.5%**	**189**	**32.9%**	**.0001**
Criterion E : Duration of the disturbance (symptoms in B, C, D)					
Duration of symptoms > 1 month	503	51.9	303	52.6%	.0001
Current presence of symptoms	350	36.6	219	38.0%	.0001
Criterion F : Dysfunctioning	**251**	**25.9%**	**103**	**18.3%**	**.0001**
Current full PTSD pattern	**54**	**6.0%**	**32**	**6.6%**	
Resolved PTSD pattern at T1	**6**	**.7%**			
Resolved PTSD pattern at T2	--	--	**14**	**3.0%**	
Delayed PTSD pattern after T1	--	--	**16**	**3.7%**	
No PTSD pattern			**444**	**90.6%**	

A total PTSD prevalence of 6% (n = 54) was found at T1, and of 6.6% (n = 32) at T2. The proportion of victims who did not show the symptoms required for a full PTSD anymore at T2 (51.9%, n = 14) was higher than the proportion of delayed onset of symptoms required for a full PTSD at T2 (4.1%, n = 18), $\chi^2(1, N = 463) = 78.86, p < .0001$.

Determinants

Degree of exposure

Most of the 54 respondents satisfying all PTSD criteria at T1, belonged to the primary victims category; i.e. 40 were direct victims or direct witnesses of the disaster; 14 were burnt due to the explosion (14/31, 45% PTSD prevalence[1]), 7 had witnessed wounded or dead victims (7/48, 15% PTSD prevalence), and 19 had witnessed the explosion but not human damage (19/532, 4% PTSD prevalence). The most afflicted indirect victims of the disaster were the family members or colleagues of someone who was on the site (11/102, 11% PTSD prevalence). Finally, 3 persons that could have been on the site but were not had also developed PTSD symptoms (3/54, 2% PTSD prevalence). None of the proxies of persons that could have been on the site met all criteria for PTSD. Prevalence and severity of the symptoms index between the primary and the secondary victims at T1, were not significantly different, $\chi^2(1, N = 510) = 2.46, p = .12$ for PTSD prevalence and $t < 1$ for PTSD severity.

These results were confirmed by a repeated measure ANOVA examining the severity of the PTSD symptoms in the different exposure groups over time. The effect of time on the mean PTSD symptoms severity was not significant, $F(1, 521) = 1.91, p = .17, \eta^2 = .00$, but the main effect of type of exposure was very significant, $F(8, 521) = 16.26, p = .000, \eta^2 = .20$. Post-hoc analyses showed that the severity of the PTSD symptoms evolved in a significant way as follows; injured and hospitalized victims showed more severity than injured but not hospitalized victims who, in their turn, showed the same symptoms severity as the witnesses of human suffering. However, the latter witnesses showed more severity than those who witnessed the explosion only, sharing the same severity in symptoms with individuals who were close to direct victims of the explosion. Individuals who could have been involved in a direct or indirect way but have not been, showed still less severity in symptoms and shared the same level of severity as other non-victims. The interaction between time and the type of exposure tended to be significant, $F(1, 521) = 1.67, p = .10, \eta^2 = .03$, which means that the intensity of the symptoms tended to evolve over time differently depending on the exposure group. Further analyses indicated that the impact decreased significantly for the witnesses of the explosion ($p = .0001$). A tendency toward significance was also found for the injured people hospitalized for more than 72 hours ($p = .08$), indicating a tendency for a decrease in PTSD severity from T1 to T2 and also for the people having someone close to them that could have been involved but was not ($p = .10$, same direction). For the other groups, intensity stayed stable over time.

1 Strictly spoken this is not PTSD prevalence given that no clinical assessment has taken place. Here, the term is to be understood as "satisfying all criteria for PTSD"

Determinants of PTSD symptoms at T1 and T2

Potential predictors at T1 were entered in a hierarchical regression model. First, demographic variables (age and sex) were entered. Thereafter were entered successively the degree of exposure; peritraumatic dissociation; social support variables; and, finally, whether or not the person had received psychological help following the disaster. Results indicated that the severity of PTSD symptoms was strongly and positively related to type of exposure, peritraumatic dissociation, and dissatisfaction with social support (Table 2). However, the social support network, i.e., the number of friends or relatives on who people can count on in case of difficulties, was not related to the severity of symptoms and Age was neither. Finally, psychological help received after the disaster was negatively associated with symptoms severity.

Table 2 Predictors of PTSD severity at 5 months (N = 700)

	β	t	Equation	Adj. R2	fjR2
Sociodemographic variables			$F(2, 697)$ = 2.452†	.004	.007
1. Family belonging					
2. Age	.006	.208			
3. Gender	.044	1.654†			
Type of exposure			$F(3, 696)$ = 49.491***	.172	.169
3. Exposure category	-.160	-5.483***			
Peritraumatic dissociation			$F(4, 695)$ = 132.209***	.429	.256
4. Peritraumatic dissociation	.491	17.273***			
Variables of social support			$F(7, 692)$ = 93.774***	.482	.055
5. Unsatisfaction with social support	.167	6.235***			
6. Number of supporting persons	-.038	-1.390			
7. Being an aid for other victims	-.127	-4.697***			
Psychological support			$F(8, 691)$ = 97.034***	.524	.042
8. Received psychological support	-.223	-7.874***			

† $p < 0,10$; * $p < 0,05$; ** $p < 0,01$; *** $p < 0,001$.

A second hierarchical regression was performed for the prediction of PTSD symptoms severity at T2 (Table 3). In this analysis, in addition to the predictors at T1, the severity of PTSD symptoms at T1 was entered too. Results revealed that only two predictors remained significant at T2. First, the higher the intensity of the symptoms present at T1, the more respondents still presented these symptoms at T2. Second, the initial dissatisfaction with the social support remained positively related to the severity of the symptoms present at T2. Thus, neither the degree of exposure, nor the peritraumatic dissociation reactions predicted the severity of the symptoms at T2. Also, the psychological help initially received did no longer negatively determine the intensity of PTSD symptoms at T2.

Table 3 Predictors of PTSD severity at 14 months (N = 384)

	β	T	Equation	Adj. R2	fj R2
Sociodemographic variables			F(2, 381) = 2.359†	.007	.012
1. Age	.046	1.570			
2. Gender	.006	.207			
Type of exposure			F(3, 380) = 20.764***	.134	.129
3. Exposure category	-.041	1.210			
Peritraumatic dissociation			F(4, 379) = 40.778***	.294	.160
4. Peritraumatic dissociation	-.014	-.363			
PTSD severity at 5 months			F(5, 378) = 160.058***	.675	.378
5. Severity of PTSD symptoms at T1	.772	18.896***			
Variables of social support at 5 months			F(8, 375) = 102.430***	.679	.007
6. Unsatisfaction with social support	.073	2.399*			
7. Number of supporting persons	-.011	-.362			
8. Being an aid for other victims	-.042	1.366			
Psychological support			F(9, 374) = 91.368***	.680	.001
9. Received psychological support	-.041	1.260			

† p < 0,10 ; * p < 0,05 ; ** p < 0,01 ; *** p < 0,001.

Discussion

The main findings on the prevalence of PTSD symptoms are twofold. At T1, 6,0% of the respondents met all criteria for an indication of a full PTSD while this prevalence became 6,6% at T2. These results are in harmony with the findings from current literature (Foa, Keane & Friedman, 2000; Heinrichs et al., 2005; Ursano, Caughey & Fullerton, 1994) but at the same time is surprising that the prevalence stays the same over time. Striking is that almost every respondent reported mental intrusions of the disaster both at T1 and T2.

The results on the recovery of 51.9% of the victims that presented a full PTSD image at T1 and did not present it anymore at T2 are in accordance with research findings (Norris, Tracy & Galea, 2009) that indicate how a majority of victims of disaster may react in a resilient way in the aftermath of extremely stressful experiences. The late-onset of PTSD symptoms at T2 is in accordance with the results of a recent study on the occurrence of mental health problems in the immediate aftermath of a fireworks disaster in Enschede (Smid, Van der Velden, Gersons & Kleber, 2012) in which 4% of the participants demonstrated late-onset PTSD. They reported high initial intrusion and avoidance and experienced progression of these symptoms. They were more likely than all other participants to use mental health services several years after the disaster. In this study, severe disaster exposure and perceived lack of social support were associated with late-onset of symptoms. This study confirms these results.

The results also show that several risk factors are associated with the severity of the PTSD symptoms. Demographic variables, particularly age and gender, did not appear to have predictive value. This is contrary to previous findings of studies (Kessler, Sonnega, Bromet, Hugher & Nelson, 1995; Norris, Tracy & Galea, 2009) which might have focused more on the impact of interpersonal trauma such as rape or assault instead of disaster.

Concerning the kind of exposure, the degree of potential life threat turned out to be an important determinant of the symptoms severity. Having been a direct witness of human damage or an indirect witness - being close to someone who died or got injured during the explosion - is a risk factor to a lesser extent.

The PTSD symptoms of the different victim categories showed the influence of the degree of exposure. In fact, 56 to 67% of the people who were injured and hospitalized presented a PTSD pattern at T1 and/or T2. Of the respondents who were directly exposed to the victims, either at the moment of the explosion, or because they were close to direct victims, 22 to 31% presented a PTSD symptoms pattern at T1 and/or T2. Finally, 99 to 100% of the respondents who could have been involved but were not exposed directly to the explosion or by the intermediate of someone who is close to them, never presented all the symptoms required for a full PTSD. These results indicate that the prevalence of PTSD symptoms developed as a function of the degree of exposure to the disaster and that this variable may serve as an important predictor of post-disaster PTSD symptoms. However, it remains less clear whether the type of victim, i.e., primary vs. secondary victims, would also be a relevant predictor of PTSD. Thus, it is necessary to examine more thoroughly the kind of exposure to the disaster.

The third factor, peritraumatic dissociation, was the most powerful PTSD predictor at T1, but the results of the second regression model showed that it had no longer the same predictive value at T2. A PTSD pattern at T1 appeared to be the biggest predictor of a PTSD pattern at T2. However, the absence of the evaluation of possible confounding variables in the relationship between peritraumatic dissociation and posttraumatic stress symptoms, as discussed in other studies (Van der Velden & Wittman, 2008), should caution against the view of peritraumatic dissociation as an independent predictor. For instance, controlling for pre-existing mental health problems could modify the current insights about peritraumatic dissociation. This is in line with other findings, which suggest that the reports of peritraumatic dissociation during or immediately after the particular event may be biased by the current psychological state of the affected individual (Candel & Merkelbach, 2004; Harvey & Bryant, 1999; Lensvelt-Mulders, Van der Hart, Van Ochten, Steele & Breeman, 2008; Marshall & Schell, 2002). Previous research (Briere, Scott & Weathers, 2005) indicates that the primary risk for PTSD is less whether one dissociates during the traumatic events, than whether such dissociation persists over time. Whereas peritraumatic dissociation ceases to predict PTSD at the multivariate level (Werner & Griffin, 2012) trauma-related persistent dissociation is a substantial predictor of PTSD. This study did not offer the possibility of gathering data on the pre-disaster mental health of participants or on persistent and generalized dissociation which continue to relate to PTSD status on the long term. It is unclear whether or not these findings are also applicable to the Ghislenghien disaster victims. Nevertheless, it is still advisable to include peritraumatic dissociation in screening measures intended to identify victims at risk for chronification of posttraumatic sequelae.

The development of PTSD symptoms was positively related to dissatisfaction with the provided social support, but not with the number of people (potentially) providing support which is in line with previous research (Drogendijk, Van der Velden, Gersons & Kleber, 2011; Norris, Baker, Murphy & Kaniasty, 2005). Psychological help received after the disaster was associated with development of a PTSD symptoms. It is not possible to provide an unambiguous interpretation for these findings because it might be that psychological help has not been useful or that this kind of help has been provided to those in the biggest need. But the latter hypothesis is not really confirmed by the results since only 60% of the respondents who developed PTSD symptoms (N = 31) have received psychological help and some of those who developed PTSD symptoms (N = 21) did not receive help while seven of these respondents declared that they had been in need for help.

Strengths and limitations

One of the positive aspects of this study is its specificity, adding knowledge to the consequences of technological or man-made disasters in which a massive explosion causes human suffering. The longitudinal character of the Ghislenghien study allows to shed another light on the predictors of mental health disturbances based on an original classification of disaster survivors. The specific instruments developed for this study may enhance systematic and comparative research on the health consequences of technological disasters in the future. However, the results also reflect the problems inherent to scientific research after a disaster. As always, there is the suddenness of the disaster and the difficulty in defining a control population. In this case, there was also a delay due to the time needed to the elaboration of partnerships, research protocols/conventions and obtaining the official authorization at various political levels. The extent of the study did not allow for clinical diagnosis of PTSD, leaving us with all the limitations and short comings of self-reports.

However, an important limitation is the low response rate of 18%, which makes the generalization and interpretation of the prevalence levels rather difficult. Although the obtained response rate is close to the expected participation of 20% in a postal survey (CROSP/IHE, 1995), the non-response could be due to three factors. First of all, people may not have been able to respond to the questionnaire because of hospitalization or recovery, physically or emotionally, while also other studies were conducted at the same time. Second, people may not have responded because they felt they were not involved in the disaster, or third, because they were disappointed about the government's management of the disaster. Finally, it is imaginable that this study may have suffered from response bias. In some cases there may have been an over or under representation due to a lack of independency between the scores of participants belonging to the same family or the same company. The dependency between different participants could possibly have affected the results (e.g.inflating correlations). The design of the study did not allow corrections for this possible dependency.

Conclusions

This study supports the evidence that the degree of exposure to a disaster, peritraumatic dissociation and the perceived lack of social support determine the development of PTSD symptoms. A second finding of this study is that early development of PTSD symptoms may lead to chronification on the long term. Besides, victims with some posttraumatic stress symptoms in the immediate aftermath of a disaster might appear resilient or adapted to the post-disaster reality, but may be confronted with an exacerbation of their symptoms on the long term (leading to a full PTSD).

Public authorities should invest in pre-disaster research planning and systematic assessment of disaster victims starting with early screening for mental health problems as first phase in a longitudinal design. Research should focus too on the possibilities of on-site psychological assistance to prevent (peritraumatic) dissociative reactions and negative emotions. Foreseeable dissatisfaction with insufficient social support over time may be a target for psychosocial disaster planning too.

Competing interests

The authors declare that they have no competing interests.

Authors' contributions

EDS, EZ, AV and HVO participated equally as the core research team in the conception and the design of the study, interpreted the results and drafted the preliminary and the end report for the Federal Public Service for Health, Food Chain Safety and Environment. HVO was central in the acquisition of funding and assured the general supervision of the core research team. EZ, EDS and AV collaborated in the collection of the data. EZ and AV statistically analyzed the data. EDS drafted the manuscript of this article. RK, OVDH and JM critically reviewed the document for intellectual content and style, and gave final approval for the version to be published.

Acknowledgements

The Federal Public Service for Health, Food Chain Safety and Environment offered support for this longitudinal study on the health impact of the Ghislenghien disaster on the affected population. Number: D/2006/2505/34.

References

American Psychiatric Association (1994). *Diagnostic and Statistical Manual for Mental Disorders (4th edition)*. Washington, DC: Author.

Bernat, J., Ronfeldt, H., Calhoun, K., Arias, I. (1998). Prevalence of traumatic events and peritraumatic predictors of posttraumatic stress symptoms in a non-clinical sample of college students. *Journal of Traumatic Stress,11(4)*,645–664.

Bonanno, G., Galea, S., Bucciarelli, A., Vlahov, D. (2006). Psychological resilience after disaster: New York City in the aftermath of the September 11th terrorist attack. *Psychological Science,17(3)*,181–186.

Brewin, C., Andrews, B., Valentine, J. (2000). Meta-analysis of risk factors for posttraumatic stress disorder in trauma-exposed adults. *Journal of Consulting and Clinical Psychology,68(5)*,748–66.

Briere, J., Scott, C., Weathers, F. (2005). Peritraumatic and persistent dissociation in the presumed etiology of PTSD. *American Journal of Psychiatry,165(12)*,2295–2301.

Candel, L., Merkelbach, H. (2004). Peritraumatic dissociation as a predictor of posttraumatic stress disorder: A critical review. *Comprehensive Psychiatry,45*,44–50.

City Health Information (2014. New York City Department of Health and Mental Hygiene, 33(1),1-8. http://www.nyc.gov/html/doh/downloads/pdf/chi/chi-33-1.pdf.

CROSP/IHE - École de santé publique (1995). Enquête de Santé par Interview: Aspects Méthodologiques. Bruxelles: Document de travail.

Dorn, T., Yzermans, J., Kerssens, J., Spreeuwenberg, P., Van der Zee, J. (2006). Disaster and subsequent healthcare utilization. A longitudinal study among victims, their family, and control subjects. *Medical Care,44(6)*,581–589.

Drogendijk, A., Van der Velden, P., Gersons, B., Kleber, R. (2011). Lack of social support among ethnic minorities after a disaster: Comparative study. *British Journal of Psychiatry,198*,317–22.

Finkenauer, C., Rimé, B. (1998). Socially shared emotional experiences vs. emotional experiences kept secret: Differential characteristics and consequences. *Journal of Social and Clinical Psychology,17*,295–318.

Foa, E., Keane, T., Friedman, M. (2000). Guidelines for treatment of PTSD. *Journal of Traumatic Stress,13(4)*,539–588.

Galea, S., Resnick, H., Ahern, J., Gold, J., Bucuvalas, M., Kilpatrick, D. (2002). Posttraumatic stress disorder in Manhattan, New York City, after the September 11th terrorist attacks. *Journal of Urban Health,79(3)*,340–353.

Galea, S., Nandi, A., Vlahov, D. (2005). The epidemiology of post-traumatic stress disorder after disasters. *Epidemiological Review,27*,78–91.

Grievink L, Van der Velden P, Yzermans CJ, Rooda J, Stellato RK. (2006). The importance of estimating selection bias on prevalence estimates shortly after disaster. *Annals of Epidemiology,16(10)*,782–786.

Guay, S., Billette, V., Marchand, A. (2006). Exploring the links between posttraumatic stress disorder and social support: Processes and potential research avenues. *Journal of Traumatic Stress,19(3)*,327–38.

Harvey, A., Bryant, R. (1999). Dissociative symptoms in acute stress disorder. *Journal of Traumatic Stress,12*,673–680.

Heinrichs, M., Wagner, D., Schoch, W., Soravia, L., Hellhammer, D., Ehlert, U. (2005). Predicting posttraumatic stress symptoms from pretraumatic risk factors: A 2-year prospective follow-up study in firefighters. *American Journal of Psychiatry,162*,2276–2286.

Hoogduin, C., Van Minnen, A., Verbraak, M., Van de Griendt, J. (2002). Klinisch beeld van de PTSS. In: A. Van Minnen & M.Verbraak (Eds.). *Psychologische interventies bij posttraumatische stress* (pp.1-14). Nijmegen: Cure & Care Publishers.

Hull, A., Alexander, D., Klein, S. (2002). Survivors of the Piper Alpha oil platform disaster: long-term follow-up study. *British Journal of Psychiatry,181*,433–438.

Kessler, R., Sonnega, A., Bromet, E., Hughes, M., Nelson, C. (1995). Posttraumatic stress disorder in the national comorbidity survey. *Archives of General Psychiatry,52(12)*,1048–1060.

Lensvelt-Mulders, G., Van der Hart, O., Van Ochten, J., Steele, K., Breeman, L. (2008). Relations among peritraumatic dissociation and posttraumatic stress: A meta-analysis. *Clinical Psychology Review,28,*1138–1151.

Luminet, O., Bouts, P., Delie, F., Mainstead, A., Rimé, B. (2000). Social sharing of emotion following exposure to a negatively valenced situation. *Cognition and Emotion,14,*661–88.

Maes, M., Delmeire, L., Schotte, C., Janca, A., Creten, T., Mylle, J., Struyf, A., Pison, G., Rousseeuw, P. (1998a). Epidemiologic and phenomenological aspects of post-traumatic stress disorder: DSM-III-R diagnosis and diagnostic criteria not validated. *Psychiatry Research,81,*179–193.

Maes, M., Delmeire, L., Schotte, C., Janca, A., Creten, T., Mylle, J., Struyf, A., Pison, G., Rousseeuw, P. (1998b). The two-factorial symptom structure of post-traumatic stress disorder: depression-avoidance and arousal-anxiety. *Psychiatry Research,81,*195–210.

Marmar, C., Weiss, D., Metzler, T., Delucchi, K. (1996). Characteristics of emergency services personnel related to peritraumatic dissociation during critical incident exposure. *American Journal of Psychiatry,153(7 Suppl),*94–102.

Marmar, C., Weiss, D., Metzler, T., Delucchi, K., Best, S., Wentworth, K. (1999). Longitudinal course and predictors of continuing distress following critical incident exposure in emergency services personnel. *Journal of Nervous and Mental Disease,187(1),*15–22.

Marshall, G., Schell, T. (2002). Reappraising the link between peritraumatic dissociation and PTSD symptom severity: Evidence from a longitudinal study of community violence survivors. *Journal of Abnormal Psychology,111,*626–636.

Mutter, A. (1998). Etude de la Validité Interne d'un Instrument d'évaluation de l'état de Stress Post-Traumatique : le QE-PTSD [Study on the Internal Consistency of a Questionnaire Evaluating Posttraumatic Stress Disorder: The QE-PTSD]. Thesis (Master). France: Université Paris VI Pierre et Marie Curie.

Miller, M. (2003). Personality and the etiology and expression of PTSD: a three factor model perpective. *Clinical Psychology Review,10(4),*373–393.

Norris, F., Baker, C., Murphy, A., Kaniasty, K. (2005). Social support mobilization and deterioration after Mexico's 1999 flood: effects of context, gender, and time. *American Journal of Community Psychology,36,*15–28.

Norris, N., Tracy, F., Galea, S. (2009). Looking for resilience: Understanding the longitudinal trajectories of responses to stress. *Social Science and Medicine,68,*2190–2198.

Ozer, E., Best, S., Lipsey, T., Weiss, D. (2003). Predictors of posttraumatic stress disorder and symptoms in adults: a meta-analysis. *Psychological Bulletin,129(1),*52–73.

Perry, P., Dean, B., Krenzelok, E. (1992). A regional poison center's experience with poisoning exposures occurring in schools. *Veterinary and Human Toxicology,34(2),*148-151.

Raphael, B., Wilson, J. (1993). Theoretical and intervention considerations in working with victims of disaster. In: B. Raphael & J.Wilson (Eds.). *International Handbook of Traumatic Stress Syndromes* (pp. 105-117). New York: Plenum.

Rimé, B., Philippot, O., Boca, S., Mesquita, B. (1992). Long lasting cognitive and social consequences of emotion: Social sharing and rumination. *European Review of Social Psychology,3*,225–258.

Smid, G., Van der Velden, P., Gersons, B., Kleber, R. (2012). Late-onset posttraumatic stress disorder following a disaster: A longitudinal study. *Psychological Trauma,4*,312–322. doi: 10.1037/a0023868.

Sydor, G., Philippot, P. (1996a). Conséquences psychologiques des massacres de 1994 au Rwanda. *Santé Mentale Québec,21*,229–246.

Sydor, G., Philippot, P. (1996b). Prévalence des symptômes de stress post-traumatique et intervention de prévention secondaire suite à une catastrophe humanitaire. *Revue Européenne de Psychologie Appliquée,46*,269–275.

Sydor, G. (1998). Questionnaire d'évaluation de l'état de Stress Post-Traumatique (QE-PTSD) [Questionnaire Evaluating Posttraumatic Stress Disorder]. Unpublished Manuscript. Belgium: Université catholique de Louvain, Louvain-la-Neuve.

Tierney, K. (2000). Controversy and consensus in disaster mental health research. *Prehospital and Disaster Medicine,15(4)*,181–187.

Ursano, R., Caughey, B., Fullerton, C. (1994). *Individual and Community Responses to Trauma and Disaster: The Structure of Human Chaos.* Cambridge: Cambridge University Press.

Ursano, R., Fullerton, C., (1997). *Trauma, Time and Recovery: Posttraumatic Stress Disorder, Acute and Long-Term Responses to Trauma and Disaster.* Washington, DC: American Psychiatric Press.

Van der Velden, P., Wittman, L. (2008). The independent predictive value of peritraumatic dissociation for PTSD symptomatology after type 1 trauma: A systemtic review of prospective studies. *Clinical Psychology Review,28(6)*,1009–1020.

Werner, K., Griffin, M. (2012). Peritraumatic and persistent dissociation as predictors of PTSD symptoms in a female cohort. *Journal of Traumatic Stress,25(4)*,401–407.

Winship, C., Mare, R. (1984). Regression Models with Ordinal Variables. *American Sociological Review,49*,512–525.

Yzermans, C., Gersons, B. (2002). The chaotic aftermath of an airplane crash in Amsterdam. In: J. Havenaar, J. Cwikel & E. Bromet (Eds.). *Toxic Turmoil: Psychological and Societal Consequences of Ecological Disasters.* New York: Kluwer Academic/Plenum Publishers.

CHAPTER 7

Does Exposure Type Impact Differentially over Time on the Development of Mental Health Disturbances After a Technological Disaster

E. De Soir, A. Versporten, E. Zech, H. Van Oyen, J. Mylle, R. Kleber & O. van der Hart

Paper published in *Archives of Public Health*, 73:20 (13 april 2015)
doi 10.1186/s13690-015-0066-z

Does exposure type impact differentially over time on the development of mental health disturbances after a technological disaster?

Erik De Soir, Ann Versporten, Emmanuelle Zech,
Herman Van Oyen, Jacques Mylle, Rolf Kleber, Onno van der Hart

Summary

This chapter is based on a longitudinal study which was conducted in order to assess the impact of the Ghislenghien disaster (July 30th, 2004) on physical, mental and social health in the affected population. The present study explored the risk for the development of four types of mental health disturbances (MHD) due to exposure to different aspects of this technological disaster in comparison with data obtained from previous health surveys among the population of the same province.

Surveys were conducted 5 months (T1) and 14 months (T2) after the disaster. Potential adult victims (≥15 years) were included (n = 1027 and 579 at T1 and T2 respectively). The "Symptom Checklist-90-Revised" (SCL-90-R) has been used in order to compute actual prevalence rates of somatization-, depression-, anxiety- and sleeping disturbances for three defined exposure categories: direct witnesses who have seen human damage (SHD), direct witnesses who have not seen human damage (NSHD) and indirect witnesses (IW). Those prevalence rates were compared with overall rates using the inhabitants of the province of Hainaut (n = 2308) as reference population. A mental health co-morbidity index was computed. Relative risks were estimated using logistic regression models.

Prevalence rates of the four MHD were much higher for the SHD than for the other exposure groups, at T1 and T2. Moreover, NSHD and IW had no increased risk to develop one of the 4 types of MHD compared to the reference population. The SHD had at T1 and T2 good 5-times a higher risk for somatization, about 4-times for depression and sleeping disorders, and 5- to 6-times for anxiety disorders respectively. Further, they suffered 13 times, respectively 17 times more from all mental disorders together.

The present study calls attention to the fact that mental health problems disturbances are significantly more prevalent and long-lasting among survivors who have directly been exposed to human damage.

Keywords: technological disaster, mental health disturbances, disaster survivors

Background

People exposed to potentially traumatic events in the context of a disaster often suffer from a wide range of psychological symptoms including anxiety-, depression-, and posttraumatic stress symptoms as well as physical symptoms such as somatization (Galea, Nandi & Vlahov, 2005; Norris et al., 2002; Versporten, De Soir, Zech & Van Oyen, 2009). Disaster survivors are often difficult to define and the denominator is usually unclear in the direct aftermath of disaster. For this reason, disaster research often focuses on the residents of the official disaster area (Yzermans & Gersons, 2002) or survivors who had to be relocated after the disaster (Grienvink et al., 2006). In other cases (Dorn et al., 2006), victims are identified through the medical chain and as a function of their medical condition. It is also well known that there exists a relationship between the severity of exposure and the mental health condition, called dose–response relationship, which could serve as a basis for victims classification. Severe exposure such as threat of life, the confrontation with injury and human losses or severe initial stress reactions may be considered as event-specific risk factors (Bonanno, Galea, Bucciarelli & Vlahov, 2006; City Health Information, 2014; Van den Berg, Grievink, Yzermans & Lebret, 2005).

On July 30th 2004, an accidental leakage in a high pressure gas pipe, which passed under the industrial zone of Ghislenghien (Belgium) exploded and instantly killed 24 people by the blast of the explosion. From the first fire crew, only two firefighters survived the initial blast and 132 people were injured. When the first crew of firefighters arrived on-scene, an enormous explosion took place: the heat of the fire was felt up to two kilometers from the explosion site. An impressive column of fire rose into the air and the heat was felt up to two kilometers away from the explosion site. Debris from the gas pipe and buildings was projected up to six kilometers away from the epicenter; up to 16 km from the explosion, air vibrations were registered. Hundreds of fire, rescue and police personnel rushed to the disaster area and all the hospitals in the region received numerous victims. A wide area was affected by the largest technological disaster that Belgium ever knew since the mine disaster of Bois du Casier in Marcinelles in which a fire caused the death of 136 Italian and 95 Belgian mineworkers (1956).

Predictors of the intensity of the PTSD symptoms among adult survivors of the Ghislenghien disaster have already been described in De Soir et al. (2008). The kind of exposure to the disaster, in particular, the degree to which life threat was experienced, was a predictor of the severity of PTSD symptoms. Survivors were classified in three main categories: victims who had been directly exposed to the disaster and direct witnesses (primary victims), indirect witnesses (secondary victims) i.e. who had been exposed to the disaster by the intermediate of the affective proximity to a victim, and, people who could have been exposed directly to the disaster (De Soir et al., 2008).

This article aims to assess the impact of the Ghislenghien gas explosion at two different points in time on the prevalence rates of four mental health disorders (MHD), namely somatization-, depression-, anxiety- and sleeping disorders, in a population affected by the disaster depending on three types of exposure. In addition, the prevalence rates will be compared with a Belgian reference population, enabling the quantification of the risk of having a MHD that can be attributed to the disaster.

Since the current literature indicates a dose–response relationship between exposure and mental health condition a classification of exposure types has been set up. Categories are have seen human damage (SHD), not have seen human damage (NSHD) and indirect witnessing (IW)

Methods

Procedures

Study population

The target group was composed of all potential victims of the Ghislenghien gas explosion. They included 1) residents of surrounding villages living up to maximum 5 km from the explosion epicenter near the industrial site and 2) employees, whether present or not, of 4 companies located on the industrial ground of Ghislenghien, as well as 3) their family members, including children from 8 to 14 years old. All the above subjects were connected with the disaster through a geographical or professional proximity as well as connections through relatives. Participation to the study was voluntary. To compare with the reference population (see below), only adult persons – i.e. aged 15 years and older - have been selected for this paper.

Study design

Aside of the classic socio-demographic data, five mental health indicators were measured at two time points. The first was in December 2004, 5 months after the disaster (T1). A follow-up assessment was carried out in September 2005, 14 months after the disaster (T2), enabling the evaluation of changes in health approximately one year after the disaster. As such, 1027 people (49.8% men, 50.2% women) agreed to participate in the study at T1 and 579 people (48.9% men, 51.1% women) at T2. The response rate at household level at T1 and T2 was 18% (n = 607 families) and 56% (n = 338 families). Ages ranged from 15 to 92 years with a mean age of 44.98 years (Se = 0.52). More details on the study design can be found in Versporten et al. (2009).

Reference population

The current study used the same mental health instruments as the 2001 Belgian National Health Interview Survey, enabling us to use those data as a reference (Bayingana et al., 2006). Ghislenghien is located in the province of Hainaut in Belgium. For reasons of closest comparability it was decided to include data of persons aged

15 years and older from the province of Hainaut (n = 2308), rather than those of the general Belgian population.

Exposure classification of victims

People described their degree of exposure to the disaster in the questionnaire by selecting among three exposure categories according to the individual's proximity to the disaster.

1. The first category encompasses the direct witnesses who have seen human damage (SHD) (n = 84). Those persons were 'active witnesses' of the disaster i.e. they witnessed the deceased or severly burnt victims, tried to offer their help and were directly exposed to effects of the explosion as they were present on site at the moment of the disaster. They were expected to be more prone to adverse health consequences due to the witnessing of grotesque scenes at the site and the life treat they have experienced (Norris, Friedman & Watson, 2002).
2. The second category are the direct witnesses who have not seen human damage (NSHD) (n = 597). They were 'passive witnesses' of the explosion such as local residents living in the surrounding communities up to 5 kilometers from the industrial site; they had heard, seen, smelled and experienced the disaster from a distance.
3. The third category concerns the indirect witnesses (IW) (n = 346). They have been indirectly exposed to the disaster through an affective proximity with a SHD or a NSHD. They were family members or colleagues of deceased or wounded persons. (e.g. co-workers not present on site, partner of a SHD).

A detailed overview of the defined categories, including their response rates at both time points, have been reported in Versporten et al. (2009).

Mental health assessment

Mental health was assessed at the T1 and T2 using the depression-, anxiety-, disturbances of sleep and somatization symptoms subscales of the "Symptom Checklist-90-Revised" (SCL-90-R) to detect cases for one or more of the above and measure their intensity (Derogatis, 1977, 1987). This self-report checklist inquires the current psychological state during the preceding week without making reference to the normal state. In this way, chronic problems are taken into account. A total score equals the sum of item scores (0 to 4), divided by the number of items in the subscale considered. Cases with more than 4 items missing on a given subscale were excluded for analysis. A threshold of 2 (SCL-score 0–1 versus 2–4) was used to assign the respondents to a group with - or a group without substantial disturbance:

- Somatization is defined by chronic complaints of widespread physical symptoms across multiple organ systems (Norris, Friedman & Watson, 2002). It refers to the development of physical symptoms for which no organic cause is found but which could have a psychological cause (Derogatis, 1977, 1987) (12-item scale);
- Depression is defined as the common concept of the 'depressive syndrome' with important characteristics as there are changes in good spirits, decline of energy, decrease of everyday activity, less capacity of oneself feeling fine, lack of interest, losing one's concentration and feeling oneself inexplicably tired (13-item scale);
- Anxiety covers an emotional component (concern, fear, terror, etc.) as well as a physical one (tensed muscles, trembling, dry mouth, sweating, stomachache, diarrhea, etc.) (10-item scale);
- Disturbances of sleep encompassed problems to fall asleep, to wake up and the quality of sleep (3-item scale).

To compare our data with existing data on mental health disorders in the reference population, a co-morbidity scale was created : 0 up to 4 mental health disorders (MHD), 1 to 4 versus 0 MHD, and 1 to 3 or 4 versus 0 MHD respectively (see below).

Data analysis

Prevalence rates of somatization-, depression-, anxiety- and sleeping disturbances as well as a mental health co-morbidity index were computed for the three defined exposure categories and compared with the reference population.

Multivariate logistic regression models and multinomial logistic regression models adjusting for age (categorized in 6 categories: 15–24, 25–34, 35–44, 45–54, 55–64, >65 years), sex and educational level (lower secondary or less education versus higher secondary education or university level) were used. Relative risks for mental health disorders, compared to the reference population, were computed. Finally, a proportional odds model was used. This model takes into account the order of the outcome variable, namely the number of MHD reported. The latter results are less precise but more concise than working at the symptom/cluster level. Computations were performed with SPSS 16.0 and STATA 8.

Results

The prevalence rates of the four examined MHD were much higher for the SHD (28 to 48 %) than for the NSHD, the IW and the reference group (10 to 20 %) at T1. Those figures remained more or less the same at T2 except a slight reduction in prevalence for depression (see Figure 1).

Figure 1. Prevalence of mental disturbances by type of exposition at 5 and 14 months after the disaster as compared with data of the province of Hainaut (Health Interview Survey 2001).

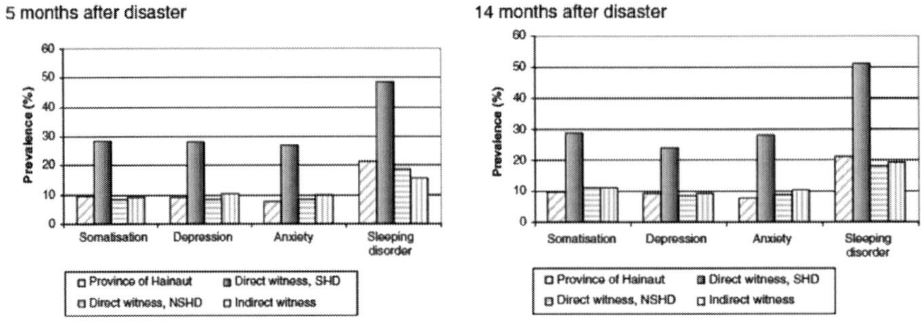

Figure 1 Prevalence of mental disturbances by type of exposition at 5 and 14 months after the disaster as compared with data of the province of Hainaut (Health Interview Survey 2001).

Logistic regressions gave evidence of 4- to 5 times higher rates among the SHD as compared to the reference population at T1 in all four MHD. Those rates slightly increased by 0.3 to 0.7 at T2 except for sleeping disorders (decrease by 0.3). Overall, NSHD and IW did not differ from the reference group at T1 nor at T2 (see Table 1).

Table 1. Odds ratios of the 4 mental health disturbances by exposure as compared to the province of Hainaut adjusted for sex, age and educational level 5 and 14 months after the disaster

	Somatization		Depression		Anxiety		Sleeping disturbance	
	OR (95%CI)		OR (95%CI)		OR (95%CI)		OR (95%CI)	
	5 months	14 months	5 months	14 months	5 months	14 months	5 months	14 months
Province Hainaut (Ref.)	1	1	1	1	1	1	1	1
Direct witness, SHD	5.2 (3.0-9.2)	5.6 (2.7-11.6)	4.3 (2.4-7.7)	4.6 (2.2-9.6)	5.0 (2.8-8.9)	5.9 (2.9-12.2)	4.6 (2.9-7.3)	4.2 (2.3-7.9)
Direct witness, NSHD	0.9 (0.6-1.4)	1.1 (0.7-1.7)	0.8 (0.6-1.2)	0.8 (0.5-1.3)	1.0 (0.7-1.5)	1.1 (0.7-1.7)	1.3 (0.9-1.6)	0.8 (0.5-1.1)
Indirect witness	1.3 (0.8-2.1)	1.7 (0.9-2.9)	1.3 (0.9-2.1)	1.2 (0.7-2.1)	1.2 (0.8-1.9)	1.6 (0.9-2.7)	1.5 (1.1-1.9)	0.9 (0.6-1.5)
Sex (women vs men)	1.2 (1.1-1.4)	1.3 (1.2-1.5)	1.1 (0.9-1.2)	1.1 (1.0-1.2)	0.9 (0.9-1.0)	0.9 (0.9-1.1)	1.2 (1.1-1.3)	1.2 (1.1-1.3)
Age (by 10 years)	2.0 (1.5-2.7)	2.0 (1.4-2.8)	2.1 (1.5-2.8)	2.3 (1.6-3.2)	2.2 (1.6-3.0)	2.1 (1.5-3.0)	1.2 (0.9-1.4)	1.5 (1.2-1.9)
Educational level (low vs high)	1.6 (1.1-2.1)	1.6 (1.1-2.2)	1.6 (1.2-2.2)	1.4 (1.0-2.0)	1.6 (1.1-2.2)	1.4 (0.9-2.0)	1.4 (1.1-1.7)	1.3 (1.0-1.7)

SHD = Seen human damage, NSHD = Not seen human damage.
Total N at 5 months = 2184 subjects (Ghislenghien n = 966; Hainaut n = 1218); Total N at 14 months = 1708 subjects (Ghislenghien n = 492; Hainaut n = 1216).
Multilogistic regression model.

The co-morbidity prevalence was much higher among SHD compared to the other groups, at T1, the most striking difference being for 4MHD (see Figure 2): Some 15.% of the SHD versus only 1.7% of the NSHD and 4.0% of the IW at T1. At T2, the are even 17.0% of the SHD with four MHD, while the prevalence in the other categories did not change. The rates of having 2 or 3 MHD decreased at T2 as compared to T1 for the SHD but remained stable for all other groups. Moreover the prevalence of having only 1 MHD increased in all victim categories at T1 but most for the SHD.

Figure 2. Prevalence of the number of mental disturbances by type of exposition at 5 and 14 months after the disaster as compared with data of the province of Hainaut (HIS 2001)

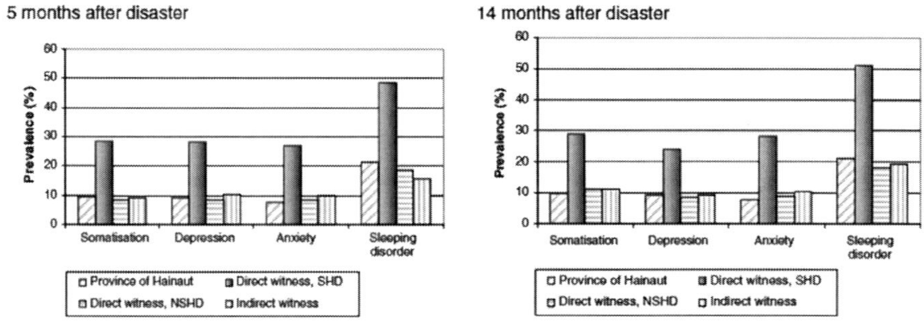

Figure 1 Prevalence of mental disturbances by type of exposition at 5 and 14 months after the disaster as compared with data of the province of Hainaut (Health Interview Survey 2001).

Table 2 presents the cumulative frequency distribution of having 1 up to 4 MHD against the prevalence of persons without any MHD. At T1, 55.2% of the SHD, 24.7% of the NSHD and 21.8% of the IW, had at least one specific recent mental problem as compared to 28.4% in the reference population. At T2, this amount remained the same for the SHD, slightly increased in the NSHD and approaches the prevalence rate of the reference population in the IW.

Table 2. Distribution of cumulative prevalence (%) of having 1 up to 4 mental disturbances versus persons without mental disturbance at 5 and 14 months after the disaster

	No mental disturbances		Having 1–4 mental disturbances	
Province of Hainaut	71.6		28.4	
	5 months		14 months	
Direct witness SHD	44.8	44.7	5 months	14 months
Direct witness NSHD	75.3	74.2	55.2	55.3
Indirect witness	78.2	72.4	24.7	25.8

SHD = Seen human damage, NSHD = Not seen human damage.

Three different models were employed to test whether the people affected by the Ghislenghien disaster had a higher risk to suffer from one or more MHD as compared to the reference population. Odds ratios were computed for the 3 exposure groups. People with no mental health disorder were taken as reference group (see Table 3). First, a multivariate logistic regression model calculated stratum specific estimates of having 1 to 4 MHD, adjusted for sex, age and educational level. The results indicated that SHD had a 3.6 times higher risk of having 1 or more MHD at T1. This ratio remained the same at T2. NSHD and IW do not differ from the reference population (OR around 1). Second, a multinomial logistic regression model revealed odds ratios of having 1 to 3 MHD respectively 4 MHD by exposure type as compared to the reference population. The OR for 4 MHD in the SHD group is 12.8 at T1 and increases to 16.9 at T2. It should however be noted that large confidence intervals were found, probably due to the limited number of cases within the SHD group (n = 84 at T1 and n = 47 at T2). The risk of having all 4 MHD in the NSHD is about 1 but about 2 in the IW. Finally, a proportional odds model allowed to estimate the odds ratio of having 1 MHD more in the victims population than the reference population. SHD had a proportional odd of about 5 at T1. In other words, the chances of going, for example, from having 2 MHD to at least 3 or 4 MHD are 5 times higher in the SHD group. Similar results were found at T2 for all groups.

Table 3. Odds ratios of having 1 up to 4 MHD (mental health disturbances); 1 up to 3 MHD / 4 MHD; and odds ratios of being in a higher numbered category (having 1 MHD more) by exposure type as compared to the province of Hainaut adjusted for sex, age and educational level at 5 and 14 months after the disaster

	1- 4 MHD[1]		1 - 3 MHD[2]		4 MHD[2]		Odds ratios of being in a higher numbered category[3]	
	OR (95% CI)		OR (95% CI)		OR (95% CI)		OR (95% CI)	
	5 months	14 months	5 months	14 months	5 months	14 months	5 months	14 months
Province Hainaut (Ref.)	1	1	1	1	1	1	1	1
Direct witness, SHD	3.6 (2.3-5.9)	3.6 (1.2-6.7)	2.8 (1.7-4.7)	2.5 (1.3-5.2)	12.8 (5.7-28.4)	16.9 (6.2-45.9)	4.7 (2.9-7.4)	4.9 (2.7-9.0)
Direct witness, NSHD	0.9 (0.7-1.1)	0.8 (0.6-1.1)	0.8 (0.7-1.1)	0.8 (0.6-1.1)	0.7 (0.3-1.4)	1.0 (0.4-2.5)	0.9 (0.7-1.1)	0.8 (0.6-1.1)
Indirect witness	0.9 (0.6-1.2)	1.1 (0.8-1.6)	0.7 (0.5-1.0)	1.0 (0.7-1.5)	1.8 (0.9-3.6)	1.8 (0.6-4.9)	0.9 (0.7-1.3)	1.2 (0.8-1.7)

	1- 4 MHD[1]		1 - 3 MHD[2]		4 MHD[2]		Odds ratios of being in a higher numbered category[3]	
	OR (95% CI)		OR (95% CI)		OR (95% CI)		OR (95% CI)	
	5 months	14 months	5 months	14 months	5 months	14 months	5 months	14 months
Sex (women vs men)	1.5 (1.2-1.8)	1.7 (1.4-2.1)	1.5 (1.2-1.8)	1.6 (1.3-2.1)	2.6 (1.5-4.6)	2.6 (1.4-5.0)	1.5 (1.2-1.8)	1.6 (1.3-2.0)
Age (by 10 years)	1.2 (1.1-1.2)	1.2 (1.1-1.3)	1.2 (1.1-1.2)	1.2 (1.1-1.3)	1.1 (0.9-1.3)	1.2 (0.9-1.4)	1.2 (1.1-1.2)	1.2 (1.1-1.3)

SHD = Seen human damage, NSHD = Not seen human damage.
1 Multivariate logistic regression model.
2 Multinomial logistic regression model.
3 Proportional odds model (ordinal cumulative logit model).
Use of weighted data for the reference population, the province of Hainaut, weight for the Ghislenghien population =1.
The reference category is 'having no MHD'.
Quest.1: Ghislenghien n = 966 (direct witness SHD : n = 79, direct witness NSHD : n = 559, indirect witness : n = 328), Province of Hainaut n = 1231.
Quest.2: Ghislenghien n = 499 (direct witness SHD : n = 43, direct witness NSHD : n = 290, indirect witness : n = 166), Province of Hainaut n = 1231.

Discussion

Our findings indicate that mental health disturbances were more prevalent in the population that was closely involved in the Ghislenghien disaster as compared to the more distant groups and to the reference population, who do not differ among each other. This is on the one hand indicative of a dose–response relationship but on the other hand of a floor effect. Moreover, the direct witnesses who had seen human damage (SHD) had significantly higher prevalence rates of recent somatization-, depression-, anxiety- and sleeping disturbances both at T1 and T2. In addition, the very high percentage of co-morbidity among the SHD is striking. Significantly more often them suffered from the four mental problems together as compared to the reference population and the other victim groups. Similar results were found at T2, suggesting that, for these victims, the impact of the disaster on mental health had not been resolved over time. This is not consistent with some studies showing a decreasing level of depression, anxiety and PTSD over time (APA, 1994; North, Kawasaki, Spitznagel & Hong, 2004) while other studies show that certain symptoms continue to exist many years after (Shore, Tatum & Vollmer, 1986).

An individual response to a disaster depends on event factors as well as individual factors. Our results confirm that physical proximity to the disaster (event factor) was a predominant criterion for the observed psychological impact (Gatchel, Schaeffer & Baum, 1985; Ursano, Fullerton, Kao & Bhartiya, 1995) because people who have been exposed to the Ghislenghien disaster in a less drastic way (NSHD and IW) did not show the above-mentioned mental health effects. Concerning the eventual effect of individual factors, we controlled for sex, age and educational level in our analyses. Our study confirms the impact of those socio-demographic variables: women are about 1.5 times more prone to adverse health effects than men, just as younger people are more affected than elder ones (OR 2) and lower educated subjects more than higher educated OR. 1.6). Furthermore we could not take into account the history or the presence of a mental disorder and any pre-existing psychopathology within the family.

Two other causes may explain the elevated prevalence rates over time. First, the way in which the Ghislenghien disaster was managed in the immediate aftermath; for example, unclear, confusing or inaccurate information was given at an early stage of the event. This may have led to increased fear, distrust, anxiety, depression and anger (Gatchel, Schaeffer & Baum, 1985; Shore, Tatum & Vollmer, 1986). However, it is known that human-made disasters often have an unclear endpoint and result in distrust of authorities (Baum & Fleming, 1993; Lonigan et al., 1994; Norwood, Ursano & Fullerton, 2000). Crisis management errors and a lack of adequate psychosocial assistance may have played a role in the detrimental mental health effects. Second, the fact that the responsibilities in this disaster were unclear for a long time may have leaded to ongoing pathogenic processes and to a long-lasting psychological impact among the SHD. Anecdotic evidence coming from personal contacts with victims during support meetings and answers on open-ended questions of the questionnaire indicated that this was particularly true for victims who reported exposure to severe human suffering.

Limitations of the study

In most traumatized populations, risk for traumatic events is confounded with pre-existing psychosocial factors (Morrow, Kamis & Hodgson, 1993). The design of our study does not rule out the possibility that some participants had histories of mental health problems before the disaster. We could not document the occurrence of symptoms and patterns of complaints prior to the disaster. Nevertheless, a causal assumption may be generated as the symptoms may be attributed to the traumatic event (Morrow, Kamis & Hodgson, 1993).

The rather low response rate of 18% at household level at T1 urges a careful interpretation of the reported prevalence rates. Indeed some response contagion may not be excluded. In addition, only 8.2% of the total study population consisted of SHD showing large confidence intervals. This suggests that other factors than the exposure type play a role in disturbing mental health. A larger number of participants would have been preferable to stabilize the estimates. The non-response could have taken place at three levels: first of all, people may not have been able to respond to the questionnaire because of hospitalization or recovery, physical or emotional. On the other hand, some may not have responded to the questionnaire because they felt they were not involved in the disaster or because they were disappointed about the government's management of the disaster or did not feel concerned by this study.

Nevertheless, this group represents about one fourth of the total number of victims that had been injured in the disaster (23.5%, 31/132 injured victims participated). Up till now, there is no systematic investigation of the mental or physical health of the population after a disaster in Belgium. If this was the case, the chances are high that more victims would accept to participate in such an investigation because there would not be the competition between various research teams. Furthermore, it is acceptable that a clear official policy followed by a widespread information campaign about the objectives of disaster research would lead to a better response rate among victims.

The previous discussion points might indicate that the prevalence rate of PTSD in victims and their families is underestimated. Due to organizational constraints of this study, the families of the deceased victims could not be contacted since their addresses have not been made available and another study on the health consequences of this disaster has been implemented in the burn injury centers. Mental health disturbances might have been more pronounced in this group.

The somatization disorder is not listed among the classic responses to a disaster, however, non-diagnostic "somatization", "somatization symptoms", and "somatic symptoms" have been abundantly described in the literature (Morrow, Kamis & Hodgson, 1993). This study used the SCL-90-R to assess people for somatic symptoms. The SCL-90-R has been the most popular symptom scale in post disaster psychiatric assessment (Norwood, Ursano & Fullerton, 2000). The scale does not provide diagnostic information, but identifies cases of somatization through the establishment of cutoff points for caseness (Bowler, Mergler, Huel & Cone, 1994). However, the adoption of symptom scales implies methodological shortcomings when used in trauma research. A first shortcoming of the SCL-90 is the construct validity whereby it fails to differentiate "somatic" symptoms (referring to any physical symptom/complaint with any medical basis) from "somatoform" symptoms (limited to physical symptoms without a medical basis). Second, somatization symptoms scales are unable to detect and correct for response sets and social desirability (Derogatis, 1987; North, 2002). Third, it has been found that the use of the SCL-90-R in trauma research measures global distress and fails to differentiate somatization from depression and anxiety (Crowne & Marlowe, 1960). These shortcomings could not be corrected in this study. Nevertheless, the 4 MHD's are interwoven. For example, anxiety can produce upset stomach, shortness of breath, sleeping difficulty, poor concentration and general agitation. Depression can also lead for example to sleep disturbances and difficulty in performing daily activities (Brophy, Norvell & Kiluk, 1988; Kass, Charles, Klein & Cohen, 1983; Kroenke et al., 1994; Kroenke, 2003). This is the reason why a co-morbidity index was developed and looked at its evolution over time.

Strengths of the study

A first strength of this study was the availability of a reference population, serving as a control group. The adults older than 15 living in the province of Hainaut were chosen as an unexposed reference population for reasons of closest comparability with the study population (Ghislenghien lies within this province). This province took part in the 2001 Belgian National Health Interview Survey. However, in spite of the use of the SCL-90-R, people may have answered differently in the aftermath of a disaster due to the social impact on the public as a whole, inducing observation bias. For example a lot of attention was given to the affected population in the media. Yet, this aspect was taken into account when collecting the second wave of data by distributing the 2nd questionnaire 14 months after the disaster instead of 12 months, being a moment in time on which a lot of attention was given in the media due to the one-year commemoration of the disaster.

This study identified a group being at particular risk for adverse health effects and may help (psychosocial) crisis managers to better tailor the resources to the needs. The people directly involved who have seen human damage (SHD) are most in need for help and this help should be provided for a longer term period as the prevalence of 4 MHD does not decrease over 14 months

Conclusions

We believe that the present study is a valuable contribution to the disaster research. The use of a reference population, serving as an ad hoc control group, and the longitudinal design enabled a scientific measurement and description of the health effects other than the classic PTSD symptoms or symptom clusters in a cause-effect perspective. These results also provided information for policy makers concerning the type and duration of health effects in disaster-affected adults and calls attention to the fact that mental health problems are long-lasting among survivors who have been witnessing human damage. We would like to advise to re-evaluate the impact of the Ghislenghien disaster on the longer term.

Competing interests

The authors declare that they have no competing interests.

Authors' contributions

EDS, EZ, AV and HVO participated equally as the core research team in the conception and the design of the study, interpreted the results and drafted the preliminary and the end reports for the Federal Public Service for Health, Food Chain Safety and Environment. HVO was central in the acquisition of funding and assured the general supervision of the core research team. EZ, EDS and AV collaborated in the collection of the data. AV statistically analyzed the data. EDS drafted the manuscript of this article and collected relevant literature references. RK, OVDH and JM critically reviewed the document for intellectual content and style, and have given final approval for this version to be published.

Acknowledgments

The Federal Public Service for Health, Food Chain Safety and Environment offered support for this longitudinal study on the health impact of the Ghislenghien disaster on the affected population (Number: D/2006/2505/3).

Consent

Written informed consent was obtained from the patient's guardian/parent/next of kin for the publication of the scientific results of the Ghislenghien study and any accompanying figures, images or data.

References

American Psychiatric Association. *Diagnostic and Statistical Manual of Mental Disorders. 4th edition.* Washington, DC: American Psychiatric Association Press; 1994.

Baum, A., Fleming, I. (1993). Implications of psychological research on stress and technological accidents. *American Psychologist,48(6)*,665–672.

Bayingana, K., Demarest, S., Gisle, L., Hesse, E., Miermans, P., Tafforeau J., Van der Heyden, J. (2006). Enquête de Santé par Interview. In: *Tableau de bord de la santé.* Santé en Hainaut,Belgique.D/2006/2505/3.http://www.hainaut.be/sante/osh/medias_user/TBS2006_CH06_Consommation_de_soins.pdf

Bonanno, G., Galea, S., Bucciarelli, A., Vlahov, D. (2006). Psychological resilience after disaster: New York City in the aftermath of the September 11th terrorist attack. *Psychological Science,17(3)*,181–186.

Bowler, R., Mergler, D., Huel, G., Cone, J. (1994). Psychological, psychosocial, and psychophysiological sequelae in a community affected by a railroad chemical disaster. *Journal of Traumatic Stress,7(4)*,601–624.

Brophy, C., Norvell, N., Kiluk, D. (1988). An examination of the factor structure and convergent and discriminant validity of the SCL-90R in an outpatient clinic population. *Journal of Personality Assessment,52(2)*,334–340.

City Health Information (2014). *New York City Department of Health and Mental Hygiene,33(1)*,1–8. http://www.nyc.gov/html/doh/downloads/pdf/chi/chi-33-1.pdf

Crowne, D., Marlowe, D. (1960) A new scale of social desirability independent of psychopathology. *Journal of Consulting Psychology,24*,349–354.

Dorn, T., Yzermans, C., Kerssens, J., Spreeuwenberg, P., Van der Zee, J. (2006). Disaster and subsequent healthcare utilization. A longitudinal studyamong victims, their family, and control subjects. *Medical Care,44(6)*,581–589.

Derogatis, L. (1977). *SCL-90-R: Administration, Scoring, and Procedures Manual, I, for the Revised Version.* Baltimore: John Hopkins University, Clinical Psychometrics Research Unit; 1977.

Derogatis, L. (1987). SCL-90-R. In: D. Kevser & D. Sweetland (Eds.). *Test Critiques Compendium: Reviews of Major Tests from the Test Critiques Series.* Kansas City, MO: Test Corporation of America.

De Soir, E., Zech, E., Versporten, A., Van Oyen, H., Kleber, R., Van der Hart, O., Mylle, J. (2008). Prédiction de l'ESPT lors d'une catastrophe technologique. *Revue Francophone du Stress et du Trauma,8(3)*,211–224.

Galea, S., Nandi, A., Vlahov, D. (2005). The epidemiology of post-traumatic stress disorder after disasters. *Epidemiologic Reviews,27*,78–91.

Gatchel, R., Schaeffer, M., Baum, A. (1985) A psychophysiological field study of stress at Three Mile Island. *Psychophysiology,22(2)*,175–181.

Grievink, L., Van derVelden, P., Yzermans, C., Roorda, J., Stellato, R. (2006). The importance of estimating selection bias on prevalence estimates shortly after disaster. *Annals of Epidemiology,16(10)*,782–786.

Kass, F., Charles, E., Klein, D., Cohen, P. (1983). Discordance between the SCL-90 and therapists' psychopathology ratings. *Archives of General Psychiatry,40(4)*,389–393.

Kroenke, K., Spitzer, R., Williams, J., Linzer, M., De Gruy, F., Hahn, S., Brody, D. (1994). Physical symptoms in primary care. Predictors of psychiatric disorders and functional impairment. *Archives of Family Medicine,3(9)*,774–779.

Kroenke, K. (2003). Patients presenting with somatic complaints: epidemiology, psychiatric comorbidity and management. *International Journal of Methods in Psychiatry Research,12(1)*,34-43.

Lonigan, C., Shannon, M., Taylor, C., Finch Jr, A., Sallee, F. (1994). Children exposed to disaster: II. Risk factors for the development of post-traumatic symptomatology. *Journal of the American Academy of Child and Adolescent Psychiatry,33(1)*,94–105.

Morrow, L., Kamis, H., Hodgson, M. (1993). Psychiatric symptomatology in persons with organic solvent exposure. *Journal of Consulting and Clinical Psychology,61(1)*,171–174.

Norris, F., Friedman, M., Watson, P., Byrne, C., Diaz, E., Kaniasty, K. (2002). 60,000 disaster victims speak: Part I. An empirical review of the empirical literature, 1981–2001. *Psychiatry,65(3)*,207–239.

Norris, F., Friedman, M., Watson, P. (2002). 60,000 disaster victims speak: Part II. Summary and implications of the disaster mental health research. *Psychiatry,65(3)*,240–260.

North, C. (2002). Somatization in survivors of catastrophic trauma: a methodological review. *Environmental Health Perspectives,110 (Suppl 4)*,637–640.

North, C., Kawasaki, A., Spitznagel, E., Hong, B. (2004). The course of PTSD, major depression, substance abuse, and somatization after a natural disaster. *Journal of Nervous and Mental Disease,192(12)*,823–829.

Norwood, A., Ursano, R., Fullerton, C. (2000). Disaster psychiatry: principles and practice. *The Psychiatric Quarterly,71(3)*,207–é26.

Shore, J., Tatum, E., Vollmer, W. (1986). Psychiatric reactions to disaster: the Mount St. Helens experience. *American Journal of Psychiatry,143(5)*,590–595.

Ursano, R., Fullerton, C., Kao, T., Bhartiya, V. (1995). Longitudinal assessment of posttraumatic stress disorder and depression after exposure to traumatic death. *Journal of Nervous and Mental Disease,183(1)*,36–42.

Van Den Berg, B., Grievink, L., Yzermans, J., Lebret, E. (2005). Medically unexplained physical symptoms in the aftermath of disasters. *Epidemiologic Reviews,27*,92–106.

Versporten, A., De Soir, E., Zech, E., Van Oyen, H. (2009). A longitudinal study on the Ghislenghien disaster in Belgium: strengths and weaknesses of the study design and influence on response rate. *Archives of Public Health,67(3)*,116–127.

Yzermans, C., Gersons, B. (2002).. The chaotic aftermath of an airplane crash in Amsterdam. In: J. Havenaar, J. Cwikel, E. Bromet (Eds.). *Toxic turmoil: psychological and societal consequences of ecological disasters.* New York: Kluwer Academic/Plenum Publishers.

CHAPTER 8

Children Following the Ghislenghien Gas Explosion: PTSD Predictors and Risk Factors

E. De Soir, E. Zech, E. Alisic, A. Versporten, H. Van Oyen, R. Kleber, O. van der Hart & J. Mylle

Paper published in *Journal of Child and Adolescent Trauma* 2014;51-62. doi 10.1007/s40653-014-0002-3

Children Following the Ghislenghien Gas Explosion: PTSD Predictors and Risk Factors

Erik De Soir, Emmanuelle Zech, Eva Alisic, Ann Versporten, Herman Van Oyen, Rolf Kleber, Onno van der Hart, Jacques Mylle

Summary

This chapter describes the risk factors for the development of posttraumatic stress reactions in children after a technological disaster in Ghislenghien, Belgium in 2004. Children were assessed at five months (T1, $N = 128$) and at fourteen months (T2; $N = 69$). At T1 and T2 respectively, 7% and 4% of the responding children showed severe posttraumatic stress reactions. Of those who showed posttraumatic stress reactions at T1, 60% recovered from these symptoms at T2. One child developed severe posttraumatic stress reactions between T1 and T2. Risk factors related to posttraumatic stress reactions were: (a) type of exposure to the disaster; (b) peritraumatic dissociation during or immediately after the disaster; and (c) dissatisfaction with the received psychological help.

Keywords: children, peritraumatic dissociation, posttraumatic stress reactions, posttraumatic stress disorder, psychological help, social support, technological disaster

Children Following the Ghislenghien Gas Explosion: PTSD Predictors and Risk Factors

On July 30th, 2004, an accidental leakage in a high-pressure natural gas pipe, that passed under the industrial zone of Ghislenghien, Belgium, created an explosion that killed 24 people and injured more than 100 people. Debris from buildings on site was projected up to six kilometers away from the epicenter. The explosion was registered up to 16 km from the explosion site. Due to the magnitude of the fire columns and the intense temperature, people burned in their cars while driving on the nearby highway. Hundreds of fire, rescue, and police personnel rushed to the disaster area and all of the regional hospitals received numerous victims. Not since the coal mine disaster in Marcinelles (August 8th, 1956) in which 278 people died had such a wide area of Belgium been affected by a large scale technological disaster.

The present study aimed at assessing the risk factors for posttraumatic stress reactions in children involved in the disaster. It examined the extent and severity of posttraumatic stress reactions in these children in regard to the type of disaster exposure and to assess the value of the psychological help that was recieved. Peritraumatic dissociation and the type of exposure were expected to be significant predictors of posttraumatic stress reactions in children.

Background

Adult survivors of disaster are typically the focus of research, but studies including child and adolescent survivors are expanding quickly. Norris et al. (2002) conducted a review of the empirical literature on disaster victims between 1981 and 2001 and found 27 relevant studies on young survivors: 17 studies concerned natural disasters, six studies involved technological disasters, and four studies focused on mass violence. Most of these studies concentrated on posttraumatic stress reactions and related disorders, but some studies focused on other problems such as health concerns and loss of resources. Fifty-two percent of the school-age samples in the review were found to suffer from severe or very severe effects. These figures indicate that children can be strongly affected by disasters. However, posttraumatic stress disrder (PTSD) rates after a disaster varied enormously in children, ranging from 0% to 95% (Saigh, Yasik, Williams, Sack, & Koplewicz, 1999). The large variability in prevalence rates may be caused by differences in age of the children, proximity to the potentially traumatizing stimuli, use of different diagnostical instruments, study design, time-lapse between exposure and assessment, and the nature of the disaster (Shaw, Applegate, & Schorr, 1996). Since Pfefferbaum's review of the literature in 1997, several authors have again noted the need for longitudinal studies on children and adolescents exposed to traumatic events, including technological disasters, because of the possibility of delayed onset or chronic persistence of PTSD symptoms (e.g., McFarlane & Van Hooff, 2009; Shah & Mulholkar, 2000).

The studies mentioned above also reported differences in factors predicting PTSD in children. For example, with regard to age, no significant differences between age groups have been found (John, Russell, & Russell, 2007). Also, with regard to gender, contradictory results have been found, although most studies reported higher symptom levels in girls (Norris et al., 2002). Other predictors of PTSD in children have been studied to a far lesser extent. Among others, a recent study on trauma exposure and PTSD in the general population (Copeland, Keeler, Angold, & Costello, 2007) found children to be more vulnerable to developing PTSD symptoms when they had suffered from an anxiety disorder prior to the exposure, when they came from an impoverished or less educated background, or when they had experienced multiple traumas. While the study of Copeland et al. (2007) took a broad approach, including psychiatric history and environmental factors, other studies have zoomed in on predictive aspects of the potentially traumatizing event itself, such as injury severity, perceived distress or negative emotions, and peritraumatic dissociation (Bryant, Salmon, Sinclair, & Davidson, 2007; Ehlers, Mayou, & Bryant, 2003). In adults, indications have been found that subjective aspects of the event (e.g., level of fear during the event) play an important role in the development of subsequent PTSD symptoms (Bernat, Ronfeldt, Calhoun, & Arias, 1998). In the aftermath of a trauma, an important challenge involves the identification of individuals who will develop PTSD. Harvey and Bryant (2002) expected the diagnosis of acute stress disorder (ASD), which differs from posttraumatic stress disorder in its requirement of three or more dissociative symptoms, to meet this challenge. The rationale is that the dissociative symptoms in the acute phase may identify those at risk of later developing PTSD. One of the most important trauma predictors appears to be peritraumatic dissociation. According to Marmar, Weiss, and Metzler (1998), peritraumatic dissociation is the occurrence of dissociative symptoms during or shortly after exposure to extreme events. It also involves acute alterations in cognitive and perceptual functioning at the time of a traumatic event.

In children, peritraumatic dissociation has hardly been studied, although some indications have been found that it is a risk factor in children as well. In a study including 45 young victims of traffic accidents, peritraumatic dissociative experiences were reported by 62% of the sample (Schäfer, Barkmann, Riedesser, & Schulte-Markwordt, 2006). Among the peritraumatic responses measured (i.e., depersonalization, derealization, and emotional numbing), derealization accounted most strongly for the variance in posttraumatic stress reactions three months after the accident, explaining 33% of the variance. Also in child survivors of sexual abuse and in children hospitalized for burns, peritraumatic dissociation has been shown to predict PTSD development (Kaplow, Dodge, Amaya-Jackson, & Saxe, 2005; Saxe et al., 2005).

Bui et al. (2010) found that the studies that examined the relationship between peritraumatic reactions and the development of subsequent PTSD symptoms in children (Dalgleish et al., 2008; Pfefferbaum et al., 2003; Schäfer et al., 2004) relied on non-validated instruments and did not discriminate between distress and dissociation. The study by Bui et al. (2010), in which children aged 8 to 15 were enrolled to participate after a traffic accident, revealed a significant association between peritraumatic variables and two measures of PTSD symptoms. The objective was to assess the predictive power of both peritraumatic distress and peritraumatic dissociation for developing acute PTSD symptoms in school-aged children. Peritraumatic distress indexes reactions such as "fear, helplessness and horror," referring to the PTSD criterion A2 in the *Diagnostic and Statistical Manual of Mental Disorders, Fourth Edition* (DSM-IV; American Psychiatric Association, 1994; Brunet et al., 2001). Peritraumatic distress appeared to be a robust predictor of who will develop PTSD symptoms among the sample.

Meiser-Stedman et al. (2007) investigated the relationships between ASD and a number of demographic, trauma, cognitive, and trauma memory variables in a sample of children and adolescents ($N = 93$) involved in assaults and motor vehicle accidents. Several cognitive variables and the quality of trauma memories, but not demographic or trauma variables, were correlated with ASD and also mediated the relationship between peritraumatic threat and ASD.

Adult research (Lensvelt-Mulders et al., 2008; Van der Hart et al., 2008) suggests that the dissociation criterion of ASD has limited validity in predicting PTSD. Dalgleish et al. (2008) addressed this issue in child and adolescent survivors ($N = 367$) of road accidents and found that dissociation accounted for no significant unique variance in later PTSD, over and above ASD criteria. Furthermore, thresholds of either three or more re-experiencing symptoms, or six or more re-experiencing/hyperarousal symptoms, were as effective at predicting PTSD as the full ASD diagnosis.

We could not locate any studies on peritraumatic dissociation in children who were victims of a disaster nor empirical evidence on the efficacy of any kind of psychological help in the immediate aftermath of a disaster. Despite this, Furr et al. (2010) conducted a meta-analytic examination of posttraumatic stress symptoms and found that disasters had a significant effect on youths' posttraumatic stress symptoms; female gender, higher death toll, child proximity, personal loss, perceived threat, and distress at the time of the event were each associated with increased posttraumatic stress symptoms. They concluded that youths are vulnerable to appreciable posttraumatic stress symptoms after disaster, with pre-existing child characteristics (i.e., gender), aspects of the disaster experience (i.e., natural vs. man-made), and study methodology each associated with variations in the magnitude of the effect.

Alisic et al. (2011) concluded, from their meta-analysis of longitudinal studies, that the most notable predictors of long-term posttraumatic stress were symptoms of acute and short-term posttraumatic stress, depression, anxiety, and parental posttraumatic stress. Female gender, injury severity, duration of hospitalization, and heart rate shortly after admission to the hospital accounted for small effects.

Method

Procedure

In the context of a general survey of people potentially involved in the accident, inhabitants living no farther than 5 km away from the explosion epicentre and employees of all companies located on the industrial site of Ghislenghien were contacted to participate in the study. Family members living at the same address and family members of all deceased persons were also invited to participate. Questionnaires were sent by mail five months (T1) after the disaster. Those who did not reply within one and a half months were contacted by telephone. Individuals agreed to participate in the study by sending a consent form or by simply filling in and sending back the questionnaire in the preaddressed envelope with stamp. Fourteen months (T2) after the disaster, each person who responded to the first questionnaire and their family members received a second questionnaire for the longitudinal follow-up of the impact of the disaster. A reminder was sent in an attempt to reach a higher response percentage at T2.

Participants

The target group of the larger study, approved by an ethical board belonging to the Scientific Institute of Public Health, comprised the survivors of the Ghislenghien gas explosion and their family members, as well as family members of deceased victims: 3,448 households, including 7,148 adults (persons aged 15 years and older on August 1st, 2004) and 820 children between the ages of 8 and 14. Children below the age of 8 years were excluded from the study sample for reliability reasons because the study included self-report data from the children's point-of-view. Five months after the disaster (T1), the response rate at household level was 18% (*N* = 607 households). Valid questionnaires were filled out by 1,027 adults (response rate of 14%) and 128 children (response rate of 16%). Fourteen months after the disaster (T2), the response rate at household level was 56% of the 607 households who received the second questionnaire. Valid questionnaires were filled out by 579 adults (46% of those who received the second questionnaire) and 69 children (53% of those who received the second questionnaire). At T1, 54.4% of the children were male and 45.6% were female. At T2, 51.4% of the children were male and 48.6% were female.

Measures

Children ranging from 8 to 14 years old received a questionnaire adapted to their age at T1. The second questionnaire (at T2) was shorter because it ommitted open questions as well as questions with invariable content such as demographics, exposure, etc.

The questionnaire contained two sections: one section with questions that had to be filled in by an adult caregiver of the child and a second section with questions that could be directly answered by the child. The first section contained questions concerning socio-demographic data, absenteeism in school, consumption of medications, utilization of health services, and both specialized medical and psychosocial care. Questions regarding recollections of the proximity and exposure to the disaster, experienced life threat, peritraumatic experiences, and posttraumatic stress reactions had to be answered by the children themselves.

Assessment of posttraumatic stress reactions. The type of disaster exposure and posttraumatic stress reactions in the children involved in the Ghislenghien disaster were assessed using a French translation of the Child Posttraumatic Stress Reaction Index (CPTS-RI; Dewulf, 2005; Frederick, Pynoos, & Nader, 1992). The CPTS-RI does not allow for a specific PTSD assessment (cf., the re-experience, avoidance, and hyperarousal symptom clusters described in the DSM-IV [American Psychiatric Association, 1994] and International Classification of Diseases-10 [ICD-10; World Health Organization, 1992]). However, this instrument allows assessment of the severity of posttraumatic stress reactions on a 5-point Likert scale ranging from 1 (*not at all or never*) to 5 (*a lot or always*). The total score is normed as follows: 12 or less indicates an absence of symptoms; between 12 and 24 refers to a pattern of mild symptoms; between 25 and 39 suggests a pattern of moderate symptoms; between 40 and 59 refers to a clinical image of severe symptoms, and more than 60 is an indication of very severe symptoms.

Type of exposure. Twenty-three yes or no questions, based on the various possible disaster-related experiences, assessed how the children were involved in or have been exposed to the disaster, allowing for a classification of the type of exposure. This classification contained three main categories which fall into seven subcategories. The first category was composed of primary victims (i.e., having been exposed to direct life threat and/or human damage and/or to the disaster itself). Three subcategories were created: (a) children who experienced direct life threat including injured children (Cat 1); (b) children who directly witnessed human suffering (burned, injured, and/or deceased victims; Cat 2); and (c) children who directly witnessed the explosion (Cat 3). The second category contained secondary victims (i.e., indirect witnesses or those who had been confronted with the disaster through the affective proximity to a primary victim). These were the family members or close relatives who either fall in subcategory four (i.e., were injured or killed in the disaster; Cat 4) or in subcategory five (i.e., who could have been injured or killed in the disaster; e.g., family members or colleagues at work who were not present at the site of the explosion; Cat 5). The third subcategory of secondary victims encompassed the children who were exposed to the disaster in a sensory way but without witnessing human damage (e.g., smell, sound, and viewing the explosion from a distance; Cat 6). Finally, the third main category, defined as tertiary victims, were children who had not been directly exposed to the disaster but were informed about it through media coverage or conversations; this class did not contain any subcategories (Cat 7). Children who did not fit in any of the previous categories were considered non-victims (Cat 8).

Peritraumatic dissociation. Reactions indicative of peritraumatic dissociation were measured by the Child Dissociation Experience Scale (Dewulf, 2005), a self-rating scale for children age 8 to 14 years, which is an adapted version of the Dissociative Experiences Scale developed for adults (Carlson & Putnam, 1993). The nine items of this scale assess sensations of emotional numbness, distancing, tunnel vision, unreality, depersonalization, out-of-body experiences, amnesia, time distortion, and anaesthesia. Children answered each question using a 5-point Likert scale ranging from 1 (*not at all*) to 5 (*to a great deal*). Since internal consistency was adequate (Cronbach's = .85), an index of peritraumatic dissociation was computed by adding the scores on the nine items. Thus, a higher total score represented a higher amount of peritraumatic dissociation.

Psychological help. The assessment of psychological help was related to the psychosocial interventions provided by psychologists or other mental health professionals and to the type (collective, individual, telephone, etc.) and number of sessions. These questions were answered by one of the parents of the child at T1. Psychological help was evaluated through the following yes or no question: "Did the child receive psychological help (personally or by telephone) after the event?" The timing of first receiving psychological help (conversation) was coded as follows: (1) the day of the event; (2) the next day; (3) during the 2 or 3 following days; (4) during the 4 or 5 following days; and (5) during the 6 or 7 following days. The amount of help received and its origin were assessed by the next two questions: "How many conversations of psychological help have there been?" and "Who took the initiative for those conversations?" The possible answers for the latter were: "I asked for help"; "I have been invited and I accepted on voluntary basis"; "I have been forced to take part in these conversations"; and "Other." Two questions related to the perceived benefits of the psychological sessions for the child were: "How much has the session been useful for the child" and "How much has the session allowed him/her to feel better?" Those questions were rated on 5-point Likert scales ranging from 1 (*not at all*) to 5 (*very much*). Finally, two yes or no questions were asked regarding the children who had not received psychological help. The first question examined whether the parent would have liked his or her child to receive psychological help and the second asked whether they had been offered help.

Results

Posttraumatic Stress Reactions in Children

The results have been interpreted using the DSM-IV-TR (American Psychiatric Association, 2000) PTSD criteria as a reference, even though the researchers did not proceed to conduct formal clinical diagnostic interviews with the children who participated in the study. As Table 1 shows, nearly all of the children who responded were exposed, to some degree, to the Ghislenghien disaster and, therefore, meet the A1 criterion (exposure to a potentially traumatic event) required for a PTSD diagnosis (American Psychiatric Association, 1994). With respect to the subjective exposure (criterion A2), at least one of the subcriteria must have been experienced in order to classify for a PTSD diagnosis. In the present sample, 75% of the children satisfy this criterion, while 64% of those children have experienced intense anxiety, 51% experienced hyperagitation at the moment of the explosion or immediately after, 50% felt powerless when the disaster occurred, and 44% reported horror.

The answers on the items indicative of symptoms belonging to the common B, C, and D clusters of the PTSD syndrome (American Psychiatric Association, 2000) were grouped in one single severity score. At T1, 21% of the children reported mild symptoms, 14% moderate symptoms, 3% severe symptoms, and 1% very severe symptoms. Conversely, this means that 61% of the sample did not report any posttraumatic stress reactions at T1. At T2, the proportion of children reporting mild symptoms was similar (21%), but the proportion of moderate to very severe symptoms had decreased over time to 4.3% and 0% respectively.

With respect to the duration (criterion E) of these symptoms, only 11% of the children reported the three types of posttraumatic stress reactions at T1, and only 6% did so at T2. This indicates that the number of posttraumatic stress reactions decreased over time.

Finally, in relation to impairment (criterion F), the disaster was found to have elicited interpersonal difficulties in 2.5% of the cases at T1 and 1% at T2. School problems, as a result of the disaster, were reported by only two children (2.5%) at both T1 and T2. Thus, it appears that the Ghislenghien disaster provoked only rather small problems in the children's social relations and at school. However, 13% of the children reported family problems at T1 and 10% at T2. At T1, 14% of the children confirmed that their disaster-related experiences made them unhappy. In sum, criterion F was endorsed by 23% of the children at T1 and 14% at T2.

Table 1. Posttraumatic Stress Reactions in Children Involved in the Ghislenghien Disaster

		T1		T2		*p* T1-T2
		N = 28	% valid	N = 69	% valid	
Criterion A1: Objective exposure	Overall	120	99.2	67	98.5	n.s.
	Cat 1: Primary victim – direct life threat	23	19.2			
	Cat 2: Primary victim – direct witness of human suffering	18	15.0			
	Cat 3: Primary victim – direct witness of the explosion (without exposure to human suffering)	29	24.2			
	Cat 4: Secondary victim – close relative injured or dead	1	.8			
	Cat 5: Secondary victim – close relative could have been injured or dead (but was not)	5	4.2			
	Cat 6: Secondary victim – partial exposure at a distance (e.g., smells, sounds, vibrations)	18	15.0			
	Cat 7: Tertiary victim – exposure via media or conversations	25	20.8			
	Cat 8: No exposure	1	.8			
Criterion A2: Subjective exposure	Overall	89	75.4	51	75.0	n.s.
	Intense anxiety	75	63.6			
	Powerlessness	59	49.6			
	Horror	52	44.1			

		T1		T2		p T1-T2
		N = 28	% valid	N = 69	% valid	
	Agitation	61	51.3			
Criterion B, C, & D: Symptoms	Overall	47	39.2	20	27.4	.0001
	No symptoms	73	60.8	53	72.6	
	Mild symptoms	25	20.8	15	20.5	
	Moderate symptoms	17	14.2	3	4.1	
	Severe symptoms	4	3.3	2	2.7	
	Very severe symptoms	1	.8	0	0	
Criterion E: Duration and current presence of symptoms	Symptoms started immediately after the disaster	15	14.0			
	Current presence of symptoms	12	11.3	4	5.6	.008
Criterion F: Dysfunction	Overall	27	22.7	10	13.7	.0001
	Disaster experience led to unhappiness	17	14.2	4	5.5	
	Difficulties in social relationships	3	2.5	1	1.4	
	Difficulties in school work	2	1.7	2	2.7	
	Difficulties in family life	16	13.2	7	9.6	
PTSR	Current PTSR	9	7.5	3	4.3	.0001
	Resolved PTSR at T1	3	2.5			
	Resolved PTSR at T2			3	4.3	
	Onset PTSR after T1			1	1.6	
	No PTSR			57	89.1	
Presence of peritraumatic dissociation during the disaster		4	3.4			

Note. T1 = Time One; T2 = Time Two; n.s. = non-significant; PTSR = Posttraumatic Stress Reactions.

The Development and Evolution of Posttraumatic Stress Reactions

Nearly half of the children (N = 4) showing severe posttraumatic stress reactions at T1 had experienced a direct threat to their life (Cat 1). Two of them developed severe posttraumatic stress reactions in the aftermath of the disaster which appeared to be resolved at T1. Of the children who reported a direct life threat at T2 (N = 23) only one suffered from severe posttraumatic stress reactions while he/she did not report these symptoms at T1. One child reporting severe posttraumatic stress reactions at T1 no longer showed these symptoms at T2. These results show that most children remained resilient over time. Two children out of the 18 children belonging to Cat 2 suffered from severe posttraumatic stress reactions at T1. These children still showed posttraumatic stress reactions at T2. One of the children presented with posttraumatic stress symptoms in the immediate aftermath of the disaster, but no longer showed these symptoms at T1 or T2. Of the children with posttraumatic stress symptoms at T1, two children out of the 28 in Cat 3 reported having been direct witnesses of the explosion, but did not see any burned or deceased victims. The 12 children who witnessed the explosion and responded at T2 did not develop severe posttraumatic stress reactions. One of these children developed posttraumatic stress reactions at T1 but did not display these symptoms at T2. The results also indicate that throughout the study 93% of the responding children did not develop severe posttraumatic stress reactions. One child, with a family member who was wounded or died (Cat 4), developed posttraumatic stress reactions at T1 but did not show a significant amount of symptoms at T2.

Degree of Exposure

Table 1 provides an overview of the children's degree of exposure to the disaster. Since none of the responding children were injured in the disaster, the most intense type of exposure experienced by the children was life threat, due to the explosion, without having been injured ($N = 23$).

In general, taking into account the norms used in the questionnaire for assessing the severity of the posttraumatic stress reactions, the results show that children who were exposed as primary or secondary victims presented with moderate to mild posttraumatic stress reactions (scores between 12 and 24). The child who had a family member that was injured or killed (Cat 4) still presented with moderate posttraumatic stress reactions at T1 (severity score between 25 and 39). The other children showed an absence of symptoms. At T2, only the Cat 1 children still presented with mild symptoms. In the other groups of children, the posttraumatic stress reactions had disappeared.

These results were confirmed through a statistical analysis (ANOVA) examining the evolution of the severity of PTSD symptoms in the different groups. The results revealed that the mean severity scores were lower at T1 than at T2, $F(1, 58) = 15.65$, $p = .0001$. The effect of the type of exposure was also significant, $F(6, 58) = 3.72$, $p = .003$. Cat 1 children showed the same amount of symptoms as the Cat 2 and Cat 3 children. These groups showed more symptoms than the Cat 5 children, whose close relative could have been injured but was not. Cat 6 and 7 children showed the least amount of symptoms. The interaction between the time of measurement and the type of exposure was not significant, $F(5, 59) = 1.36$, $p = .25$.

PTSD Risk Factors at T1 and T2

Table 2 and Table 3 show the correlations between the potential predictors, such as the degree of exposure and the severity of posttraumatic stress reactions at T1 and T2, respectively. The results of the regression analysis in Table 4 indicate that the severity of posttraumatic stress reactions at T1 was strongly related to the type of exposure (from primary to tertiary victims), peritraumatic dissociative reactions, and to received psychological help. Age and gender were not found to be linked to symptom severity. The regression model explains 42% of the total variance.

At T2, the severity of PTSD-related symptoms was best predicted by peritraumatic dissociative reactions and the severity of symptoms at T1 (see Table 5). This regression model explains 60% of the total variance.

Table 2. Correlations Between the Main Variables for the Children Included at T1

	1	2	3	4	5
1. Severity of PTSD at 5 months	--				
2. Age	-.039	--			
3. Gender	.079	.159*	--		
4. Type of exposure	-.393***	.008	.065	--	
5. Severity of dissociation	.488***	.136†	-.061	-.308***	--
6. Psychological support received	.394***	-.154†	.220**	-.191*	-.048

Note. T1 = Time One; PTSD = Posttraumatic Stress Disorder; *N* = 112; Gender (0 = boy; 1 = girl); Type of exposure (higher scores indicate lower degree of exposure); Psychological support received (0 = no; 1 = yes).
† $p < 0.10$; * $p < 0.05$; ** $p < 0.01$; *** $p < 0.0001$.

Table 3. Correlations Between the Main Variables for the Children Included at T2

	1	2	3	4	5	6
1. Severity of PTSD at 14 months	--					
2. Age	-.207†	--				
3. Gender	.230*	.024	--			
4. Type of exposure	-.442***	.060	.006	--		
5. Severity of dissociation	.501***	.106	.020	-.407***	--	
6. Severity of PTSD at 5 months	.742***	-.203†	.198†	-.477***	.458***	--
7. Psychological support received	.358***	-.229*	.205†	-.202†	-.067	.306**

Note. T2 = Time Two; PTSD = Posttraumatic Stress Disorder; *N* = 62.
† $p < 0.10$; * $p < 0.05$; ** $p < 0.01$; *** $p < 0.0001$.

Table 4. Predictors of the Severity of PTSD Symptoms at 5 Months

Severity of PTSD Symptoms at 5 Months	Std. β	*t*	Equation	Adj. R^2	ΔR^2
Model 1: Sociodemographic Variables			$F(2, 109) = .490$	-.009	.009
Age	-.053	-.544			
Gender (0 = male; 1 = female)	.087	.903			
Model 2: Exposure Type			$F(3, 108) = 7.277$***	.145	.159
Age	-.053	-.602			
Gender (0 = male; 1 = female)	.114	1.274			
Exposure (from primary victim to no victim)	-.400	-4.547***			
Model 3: Peritraumatic Dissociation			$F(4, 107) = 13.235$***	.306	.163
Age	-.117	-1.446			
Gender (0 = male; 1 = female)	.141	1.755†			
Exposure (from primary victim to no victim)	-.269	-3.226**			
Peritraumatic dissociation	.430	5.103***			
Model 4: Psychological Help Received			$F(5, 106) = 17.001$***	.419	.114
Age	-.051	-.677			
Gender (0 = male; 1 = female)	.047	.617			
Exposure (from primary victim to no victim)	-.185	-2.363*			
Peritraumatic dissociation	.458	5.929***			
Received psychological support (0 = no; 1 = yes)	.363	4.667***			

Note. PTSD = Posttraumatic Stress Disorder; $N = 112$.
†$p < 0.10$; *$p < 0.05$; **$p < 0.01$; ***$p < 0.001$.

Table 5. Predictors of the Severity of PTSD Symptoms at 14 Months

Severity of PTSD Symptoms at 14 months	Std. β	*t*	Equation	Adj. R^2	ΔR^2
Model 1: Sociodemographic Variables			$F(2, 59) = 3.202$*	.067	.098
Age	-.213	-1.718†			
Gender (0 = male; 1 = female)	.235	1.899†			
Model 2: Exposure Type			$F(3, 58) = 8.562$***	.271	.209
Age	-.240	-2.193*			
Gender (0 = male; 1 = female)	.238	2.180*			
Exposure (from primary victim to no victim)	-.458	-4.182***			
Model 3: Peritraumatic Dissociation			$F(4, 57) = 11.314$***	.403	.136
Age	-.273	-2.746**			
Gender (0 = male; 1 = female)	.230	2.324*			
Exposure (from primary victim to no victim)	-.295	-2.727**			
Peritraumatic dissociation	.405	3.724***			
Model 4: Psychological Help Received			$F(5, 56) = 17.746$***	.579	.171
Age	-.130	-1.466			
Gender (0 = male; 1 = female)	.122	1.414			
Exposure (from primary victim to no victim)	-.102	-1.026			
Peritraumatic dissociation	.222	2.258*			
Severity of PTSD at 5 months	.542	4.968***			
Model 5: Psychological Help Received			$F(6, 55) = 15.952$***	.595	.022
Age	-.102	-1.165			
Gender (0 = male; 1 = female)	.095	1.106			
Exposure (from primary victim to no victim)	-.070	-.714			
Peritraumatic dissociation	.264	2.663**			
Severity of PTSD at 5 months	.497	4.537***			
Received psychological support (0 = no; 1 = yes)	.167	1.821†			

Note. PTSD = Posttraumatic Stress Disorder; $N = 62$.
†$p < 0.10$; *$p < 0.05$; **$p <$

Psychological Help

At T1, five children (4.2%) had received psychological help in the form of a personal contact. Two of those children belonged to Cat 1, two others to Cat 2, and one to Cat 3. No one in the other categories received psychological help.

Only three of the nine children with severe posttraumatic stress reactions at T1 received psychological help. Of the six children with severe posttraumatic stress reactions at T1 who did not receive psychological help, three belonged to Cat 1, two to Cat 2, and one to Cat 3. The parents of two Cat 1 children, who developed severe posttraumatic stress reactions at T1, had expected and hoped to receive some kind of support or help, but ultimately they did not receive any help.

Type, Timing, and Perceived Benefits of Psychological Help

The timing, frequency, and kind of psychological help that children received after the disaster was diverse. Some children received psychological help three days after the disaster. Two children attended a single session, two other children received two sessions, and one child attended 10 sessions. These sessions were both individual and collective: (a) in the context of a family session ($N = 1$); (b) in the form of a telephone contact ($N = 1$); and (c) in the context of a session with someone from outside the family ($N = 1$).

Of the five children who received psychological help between the time of the disaster and T1, three developed a significant amount of posttraumatic stress reactions at T1. There was an equal proportion of each possible type of primary exposure. Between T1 and T2, two children continued to receive psychological help (five and 15 sessions, respectively). The child who received more intensive psychological help still showed a significant amount of posttraumatic stress reactions at T2, while the other child did not show any posttraumatic stress reactions at T1 or T2. Due to the small number of children who received psychological help and the diversity of interventions, it was impossible to analyze the efficacy of the interventions.

However, it was possible to examine the beneficial effects of treatment, as perceived by the parents, at T1. The parents of the children who received psychological help estimated this help as very useful for their child (four out of five parents reported the highest possible score of usefulness) and reported that the help had allowed their child to feel better or relieved (the same four out of five parents reported the highest degree of satisfaction). In just one case, the psychological help was seen as unbeneficial for the child. This child developed severe posttraumatic stress reactions at T1, but no information was available at T2 because the second questionnaire was not answered.

Discussion

This study examined risk factors influencing the severity of posttraumatic stress reactions in children involved in a technological disaster (i.e., the type of exposure to the disaster, peritraumatic dissociation, and the type of psychological help that was received).

Prevalence and Prediction of Posttraumatic Stress Reactions

The results of this study show that in the sample of responding children between the ages of 8 and 14 years, 7.5% presented severe posttraumatic stress reactions at T1 and 4.3% still showed these reactions at T2. For these children, 60% of those who presented with posttraumatic stress reactions at T1 did not present with these symptoms at T2 (three out of five responding children at T2).

These results indicate that several risk factors were associated with the severity of posttraumatic stress reactions. The first risk factor was the type of exposure, which referred to the subjective experiences of the children. Perceived life threat, experienced by the children who were primary victims, was the highest level of exposure. The secondary victims, who were direct witnesses of severe human suffering (e.g., burned, injured, and/or deceased victims), family members or close relatives of someone who was injured or died during the disaster, or direct witnesses of the explosion, were also affected, but to a lesser degree. Children who were exposed to the disaster in a sensory way (e.g., smells, sounds, and viewing the explosion) were still less affected. It might be surprising that children only showed moderate to mild posttraumatic stress symptoms after being directly confronted with life threat and/or human suffering or bereavement. While the type of exposure was expected to lead to a different subjective experience, this experience may also have been different in children exposed to the same objective facts.

The second risk factor that came out of this study was peritraumatic dissociation during or immediately after the disaster, i.e., dissociative reactions such as out of body experiences, amnesia, automatic pilot behavior, tunnel vision, disorientation, and depersonalization. Reactions of peritraumatic dissociation appeared to be the most important factor in the development of posttraumatic stress symptoms at T1. This finding is in accordance with Ozer et al. (2003) and Lensvelt-Mulders et al. (2008), but to our knowledge there are no other studies with which our findings on the impact of peritraumatic dissociation on children's disaster experiences can be compared. In a meta-analysis on predictors of PTSD in adults, one subjective response during a potentially traumatizing event stood out among risk factors for PTSD: peritraumatic dissociation (Ozer, Best, Lipsey, & Weiss, 2003). A more recent meta-analysis of studies published between 1995 and 2004 (Breh & Seidler, 2007) confirmed Ozer et al.'s results (2003), but it is not clear whether or not these findings are valid in respect to children. According to Van der Velden and Wittman (2008), a limitation in most of these studies is the absence of the evaluation of possible confounding variables in the relationship between peritraumatic dissociation and posttraumatic stress reactions. For instance, controlling for mental health problems occurring in the first few days or weeks post-event could modify the current insights about peritraumatic dissociation. This is in line with other findings which suggest that the reports of peritraumatic dissociation during or immediately after the particular event may be biased by the current psychological state of the affected individual (Bryant, 2007; Candel & Merkelbach, 2004; Harvey & Bryant, 1999; Marshall & Schell, 2002). Since the Ghislenghien study did not offer the possibility of gathering data on the pre-disaster mental health of the participants, it is unclear whether or not these findings are also applicable to children. According to Bui et al. (2010), it is possible that peritraumatic dissociation predicts the persistence of PTSD rather than its development. Our results cannot verify that, but future research could measure the peritraumatic reactions more immediately after the explosion. Although early posttraumatic stress symptoms are robust predictors of later posttraumatic stress symptoms, measuring peritraumatic distress in the children belonging to different victim categories might also have been extremely useful in predicting the development of chronic posttraumatic symptoms. Another point of discussion concerns the fact that the questions regarding the proximity and exposure to the disaster, experienced life threat, peritraumatic experiences, and posttraumatic stress reactions relied on recollections reported by the children themselves. These retrospective data could imply a memory bias in these children but a lack of empirical evidence makes it impossible to draw conclusions on this subject. Future research could benefit from collecting data from younger children as their greater dependence on adults may provide additional useful insights to the investigation of trauma associated with disaster exposure, however this was not possible in this study.

Nevertheless, it is advisable to include peritraumatic dissociation and/or distress in screening measures intended to identify children at risk for the development of chronic posttraumatic sequelae. In an explorative study with survivors of severe motor vehicle accidents, De Soir and Goffings (2012) propose utilizing a set of grounding and psychological stabilization techniques in order to counter peritraumatic dissociative states. Explorative quantitative and qualitative research on this kind of advanced psychological support indicates beneficial effects on preventing the development of posttraumatic stress reactions in accident survivors, but this still needs to be confirmed by randomized controlled trials with disaster survivors and children, more specifically.

The results mentioned above indicate that, even if the numbers are small, the children in our study should have received more help in the immediate aftermath of the disaster. Although, it is important to note that it was the parents who answered the questions regarding the psychological help that their children received and the perceived benefits of that help. This shows a possible mediating relationship between the impact of the disaster on adults and the well-being of their children. It remains unclear whether or not the provided help reached its targets.

Limitations

Although this study followed children on a longitudinal basis, results also reflect the problems inherent to retrospective research after a technological disaster. Among others, it did not take into account the psychiatric history nor the environmental factors which might have had an effect on the disaster experiences of the involved children. As always, disasters are sudden and there is difficulty in creating a control group.

Another limitation is the low overall response rate of 18% and 16% for the children. This response rate makes generalization and interpretation of the prevalence levels quite difficult. The study population included all people who could possibly have been affected. No exclusion criteria were established. The overall response rate at household level at 5 months was 18%. In order to increase the response rate among the direct victims, they were contacted by psychologists who tried to motivate them to participate in the study. A majority of the people who declined to participate did not want to be confronted with the disaster again. Furthermore, bureaucratic and political issues regarding the responsibility of a disaster situation often lead to a delay in the onset of scientific study. The low response rate necessitates careful interpretation of subsequent results. Differences in respondent characteristics may lead to a bias in prevalence estimates and associations. In order to increase the participation rate in disaster research, questionnaires could be completed during a face-to-face interview, but this implies more resources. A higher response rate in other studies could also be a result of the shorter time interval between the disaster and the first administration of measures.

There could be three different causes for the lack of response to the present study. First, people might not have been able to answer the questionnaire due to hospitalization or recovery, be it physical or emotional. These people may not have wanted to be confronted with the disaster again. Secondly, it is possible that those living farther away from the epicenter of the disaster have not filled in the questionnaires because they did not feel involved or concerned with the disaster. This could lead to a non-response bias related to the exposure. Finally, this study was not integrated in the global management of the disaster. Several official institutions approached the affected victims and provided numerous official forms to complete. This may have led to disappointment in the authorities and lack of willingness to participate in a scientific study.

The suddenness of a technological disaster puts time pressure on the researchers regarding the development of an adequate research protocol. A methodological problem concerns the definition of "affected" population. In addition to the direct victims who were on the site of the disaster, a rather large group of residents were included in the study population in order to not miss any potentially affected victim. Nevertheless, a response percentage of 20%, which was almost reached, is considered normal for postal questionnaires (CROSP/IHE, 1995).

Recommendations for Future Research

In the future, a specific monitoring system for general practitioners could be set up in order to allow for more detailed monitoring of children involved in a disaster. The results of this study indicate the variety in both the evolution of posttraumatic symptomatology and the need for psychological help. A personalized contact would raise the probability of a better diagnosis, an identification of the needs of victims, and a more valid diagnosis of posttraumatic sequelae. This would increase the quality of trauma research in groups of children involved in disasters, allowing for the development of more specific instruments to identify posttraumatic stress reactions in children involved in technological disaster.

Authors' contributions

EDS, EZ, AV and HVO participated equally as the core research team in the conception and the design of the study, interpreted the results and drafted the preliminary and the end reports for the Federal Public Service for Health, Food Chain Safety and Environment. HVO was central in the acquisition of funding and assured the general supervision of the core research team. EZ, EDS and AV collaborated in the collection of the data. AV statistically analyzed the data. EDS drafted the manuscript of this article and collected relevant literature references with the help of EA. RK, OVDH and JM critically reviewed the document for intellectual content and style, and have given final approval for this version to be published.

References

Alisic, E., Jongmans, M.J., Van Wesel, F., & Kleber, R. (2011). Building child trauma theory from longitudinal studies: A meta-analysis. *Clinical Psychology Review, 31(5)*, 736-747. doi: 10.1016/j.cpr.2011.03.001.

American Psychiatric Association. (1994). *Diagnostic and statistical manual for mental disorders, 4th Edition*. Washington, DC: APA.

American Psychiatric Association (2000). *Diagnostic and statistical manual of mental disorders (4th ed. Text edition)*. Washington DC: Author.

Bernat, J.A., Ronfeldt, H.M., Calhoun, K.S., & Arias, I. (1998). Prevalence of traumatic events and Peritraumatic predictors of posttraumatic stress symptoms in a non-clinical sample of college students. *Journal of Traumatic Stress, 11(4)*,645-664. doi: 10.1023/A:1024485130934.

Breh, D. C., & Seidler, G. H. (2007). Is peritraumatic dissociation a risk factor for PTSD. *Journal of Trauma and Dissociation, 8(1)*, 23-69. doi: 10.1300/J229v08n01_04.

Brunet, A., Weiss, D.S., Metzler, T.J., Best, S.R., Neylan, T.C., Rogers, C. Fagan, J., & Marmar, C.R. (2001). The peritraumatic distress inventory: A proposed measure of PTSD criterion A2. *American Journal of Psychiatry, 158(9)*, 1480-1485.

Bryant, R. A. (2007). Does dissociation further our understanding of PTSD? *Journal of Anxiety Disorders, 21*, 183-191. doi: 10.1016/j.janxdis.2006.09.012.

Bryant, R.A., Salmon, K., Sinclair, E., & Davidson, P. (2007). A prospective study of appraisals in childhood posttraumatic stress disorder. *Behaviour Research and Therapy, 45 (10)*, 2502-7. doi: 10.1016/j.brat.2007.04.009.

Bui, E., Brunet, A., Allenou, C., Camassel, C., Raynaud, J.-P., Claudet, I., Fries, F., Cahuzac, J.-P., Grandjean, H., Schmitt, L., & Birmes, P. (2010). Peritraumatic reactions and posttraumatic stress symptoms in school-aged children victims of road traffic accident. *General Hospital Psychiatry, 32*, 330-333. doi: 10.1016/j.genhosppsych.2010.01.014.

Candel, L., & Merkelbach, H. (2004). Peritraumatic dissociation as a predictor of posttraumatic stress disorder : A critical review. *Comprehensive Psychiatry, 45*, 44-50. doi: 10.1016/j.comppsych.2003.09.012.

Carlson, E. B., & Putnam, F. W. (1993). An update on the Dissociative Experiences Scale. *Dissociation: Progress in the Dissociative Disorders, 6*, 16–27.

Copeland, W. E., Keeler, G., Angold, A., & Costello, E. J. (2007). Traumatic events and posttraumatic stress in childhood. *Archives of General Psychiatry, 64*, 577-584.

CROSP/IHE (1995). Enquête de santé par interview: aspects méthodologiques. Document de travail, Ecole de Santé Publique (Eds), ULB.

Dalgleish, T., Meiser-Stedman, R., Kassem-Adams, N., Ehlers, A., Winston, F., Smith, P., Bryant, B., Mayou, R.A., & Yule, W. (2008). Predictive validity of acute stress disorder in children and adolescents. *The British Journal of Psychiatry, 192*, 392-393. doi: 10.1192/bjp.bp.107.040451.

De Soir, E., & Goffings, K. (2012). Directe opvang van verkeersslachtoffers. [Direct care of victims of traffic accidents] In: E. Vermetten, R.J. Kleber, & O. van der Hart (Eds), *Handboek Posttraumatische stressstoornissen* (pp. 321-335). Utrecht: De Tijdstroom.

Dewulf, A. C. (2005). Impact of accidental injuries on children and their parents: traumatic stress, social sharing and emotions. *Unpublished doctoral thesis.* Université Catholique de Louvain, Belgium.

Ehlers, A., Mayou, R.A., & Bryant, RA.. (1998). Psychological predictors of chronic posttraumatic stress disorder after motor vehicle accidents. *Journal of Abnormal Psychology, 107(3)*, 508-519. doi: 10.1037/0021-843X.107.3.508.

Ehlers, A., Mayou, R.A., & Bryant, R.A. (2003). Cognitive predictors of posttraumatic stress disorder in children: results of a prospective longitudinal study. *Behaviour Research and Therapy, 41 (1)*, 1-10. doi:10.1016/S0005-7967(01)00126-7.

Frederick, C. J., Pynoos, R. S., & Nader, K. (1992). The Child Post-Traumatic Stress Disorder Reaction Index. In K. Nader (Ed.), Childhood trauma: A manual and questionnaires. *Unpublished manuscript*. Copyrighted instrument.

Furr, J.M., Comer, J.S., Edmunds, J.M., & Kendall, P.C. (2010). Disasters and youth: A meta-analytic examination of posttraumatic stress. *Journal of Consulting and Clinical Psychology, 78(6)*, 765-780. doi: 10.1037/a0021482.

Harvey, A.G., & Bryant, R.A. (1999). Dissociative symptoms in acute stress disorder. *Journal of Traumatic Stress, 12*, 673-680. doi: 10.1023/A:1024773202939.

Harvey, A.G., & Bryant, R.A. (2002). Acute stress disorder: A synthesis and critique. *Psychological Bulletin, 128*, 886-902. doi: 10.1037/0033-2909.128.

John, P. B., Russell, S., & Russell, P. S. S. (2007). The prevalence of posttraumatic stress disorder among children and adolescents affected by tsunami disaster in Tamil Nadu. *Disaster Management and Response, 5*, 3-7. doi:10.1016/j.dmr.2006.11.001.

Kaplow, J. B., Dodge, K. A., Amaya-Jackson, L., Saxe, G. N. (2005). Pathways to PTSD, part II: Sexually abused children. *American Journal of Psychiatry, 162*, 1305-1310. doi: 10.1176/appi.ajp.162.7.1305.

Lensvelt-Mulders, G., Van der Hart, O., Van Ochten, J.M., Steele, K, & Breeman, L. (2008). Relations among peritraumatic dissociation and posttraumatic stress: A meta-analysis. *Clinical Psychology Review, 28*, 1138-1151. doi:10.1016/j.cpr.2008.03.006.

Marmar, C. R., Weiss, D. S., & Metzler, T. J. (1998). Peritraumatic dissociation and posttraumatic stress disorder. In J. D. Bremner & C. R. Marmar (Eds.), *Trauma, memory, and dissociation* (pp. 229-252). Washington, DC: American Psychiatric Press.

Marmar, C. R., Weiss, D. S., Metzler, T. J., & Delucchi, K. (1996). Characteristics of emergency services personnel related to peritraumatic dissociation during critical incident exposure. *American Journal of Psychiatry, 153(7 Suppl.)*, 94-102.

Marmar, C. R., Weiss, D. S., Metzler, T. J., Delucchi, K. L., Best, S. R., & Wentworth, K. A. (1999). Longitudinal course and predictors of continuing distress following critical incident exposure in emergency services personnel. *Journal of Nervous and Mental Disease, 187*, 15-22.

Marschall, G.N., & Schell, T.L. (2002). Reappraising the link between peritraumatic dissociation and PTSD symptom severity: Evidence from a longitudinal study of community violence survivors. *Journal of Abnormal Psychology, 111*, 626-636. doi: 10.1037//0021-843X.111.4.626.

McFarlane, A. C., & Van Hooff, M. (2009). Impact of childhood exposure to a natural disaster on adult mental health: 20-year longitudinal follow-up study. *British Journal of Psychiatry, 195(2)*, 142-148. doi: 10.1192/bjp.bp.108.054270.

Meiser-Stedman, R., Smith, P., Yule, W., Dalgleish, T., & Glucksman, E. (2007). Diagnostic, demographic, memory quality, and cognitive variables associated with acute stress disorder in children and adolescents. *Journal of Abnormal Psychology, 116(1)*, 65-79. doi: 10.1037/0021-843X.116.1.65.

Nijenhuis, E. R., Van Dyck, R., Spinhoven, P., Van der Hart, O., Chatrou, M., & Vanderlinden, J. (1999). Somatoform dissociation discriminates among diagnostic categories over and above general psychopathology. *Australian and New Zealand Journal of Psychiatry, 33*, 511-520. doi: 10.1046/j.1440-1614.1999.00601.x.

Norris, F. H., Friedman, M. J., Watson, P. J., Byrne, C. M., Diaz, E., & Kaniasty, K. (2002). 60,000 disaster victims speak, Part I: An empirical review of the empirical literature, 1981-2001. *Psychiatry, 65*, 207-239. doi: 10.1521/psyc.65.3.207.20173.

Norris, F. H., & Kaniasty, K. (1992). A longitudinal study of the effects of various crime prevention strategies on criminal victimization, fear of crime, and psychological distress. *American Journal of Community Psychology, 20*, 625-648. doi: 10.1007/BF00941775.

Ozer, E. J., Best, S. R., Lipsey, T. L., & Weiss, D. S. (2003). Predictors of posttraumatic stress disorder and symptoms in adults: A meta-analysis. *Psychological Bulletin, 129*, 52-73. doi: 10.1037/0033-2909.129.1.52.

Pfefferbaum, B. (1997). Posttraumatic stress disorder in children: A review of the past 10 years. *Journal of the American Academy of Child & Adolescent Psychiatry, 36(11)*, 1503-1511.

Pfefferbaum, B., North, C.S., Doughty, D.E., Gurwitch, R.H., Fullerton, C.S., & Kuyla, J. (2003). Posttraumatic stress and functional impairment in Kenyan children following the 1998 American Embassy bombing. *American Journal of Orthopsychiatry, 73(2)*, 133-140. doi: 10.1037/0002-9432.73.2.133.

Saigh, P. A., Yasik, A. E., Williams, H., Sack, M. D., & Koplewicz, M. D. (1999). Child-adolescent posttraumatic stress disorder: Prevalence, risk factors, and comorbidity. In P.A. Saigh & J.D. Bremner (Eds.), *Posttraumatic stress disorder: A comprehensive text* (pp. 18-43). Boston: Allyn & Bacon.

Saxe, G. N., Stoddard, F., Hall, E., Chawla, N., Lopez, C., Sheridan, R., King, D., King, L, & Yehuda, R. (2005). Pathways to PTSD, Part I: Children with burns. *American Journal of Psychiatry, 162*, 1299-1304. doi: 10.1176/appi.ajp.162.7.1299.

Shäfer, I., Barkmann, C., Riedesser, P. & Schulte-Markwort, M. (2004). Peritraumatic dissociation predicts posttraumatic stress in children and adolescents following road traffic accidents. *Journal of Trauma & Dissociation, 5(4)*, 79-90. doi: 10.1300/J229v05n04_05.

Shah, N., & Mudholkar, S. (2000). Clinical aspects of post-traumatic stress disorder in children and adolescents. In K. N. Dwivedi (Ed.), *Post-traumatic stress disorder in children and adolescents* (pp. 97-112). Philadelphia, PA US: Whurr Publishers.

Shaw, J. A., Applegate, B., & Schorr, C. (1996). Twenty-one-month follow-up study of school-age children exposed to Hurricane Andrew. *Journal of the American Academy of Child and Adolescent Psychiatry, 35,* 359-364. doi: 10.1097/00004583-199603000-00018.

Van der Hart, O., Van Ochten, J., Van Son, M.J.M., Steele, K., & Lensvelt-Mulders, G. (2008). Relations among peritraumatic dissociation and posttraumatic stress: A critical review. *Journal of Trauma & Dissociation, 9(4),* 481-505.

Van der Velden, P. G., & Wittman, L. (2008). The independent predictive value of peritraumatic dissociation for PTSD symptomatology after type 1 trauma: A systematic review of prospective studies, *Clinical Psychology Review, 28,* 1009-1020. doi: 10.1016/j.cpr.2008.02.006.

World Health Organization (1992). *The ICD-10 Classification of Mental and Behavioural Disorders: Clinical Descriptions and Diagnostic Guidelines.* Geneva: Author.

Zatzick, D. F., Grossman, D. C., Russo, J., Pynoos, R., Berliner, L., Jurkovich, G., Sabin, J.A., Katon, W., Ghesquiere, A., McCauley,. & Rivara, F. P. (2006). Predicting posttraumatic stress reactions longitudinally in a representative sample of hospitalized injured adolescents. *Journal of the American Academy of Child and Adolescent Psychiatry, 45,* 1188-1195. doi:10.1097/01.chi.0000231975.21096.45.

CHAPTER 9

Discussion and Conclusions

Discussion and Conclusions

9.1. Preliminary

The aim of this thesis was to examine the impact of a technological disaster, both phenomenologically and empirically. The rationale of this research has been four-fold: 1) the lack of knowledge concerning the mental health disturbances after a technological disaster originated by (an accidental) massive explosion; 2) the diversity of PTSD prevalence in the literature on adult and child disaster survivors; 3) the current discussion on the determinants of posttraumatic stress symptoms, and more specifically the influence of peritraumatic dissociation, exposure to the disaster and social support as risk factors; and, 4) the relative lack of attention given to the conceptual differences between the concept of psychological trauma used in the Anglo-Saxon literature and the more hermeneutic trauma interpretations in French theory.

9.2. Statements and discussion on the mental health disturbances in the wake of technological disaster

9.2.1. The concept of psychological trauma in French contemporary theory is based on clinically useful concepts: 'syndrome de répétition', 'effroi de la mort 'and 'névrose traumatique'.

The concept of psychological trauma in contemporary French trauma theory has been inspired by the classical Freudian view on overwhelming events. The empirically focused Anglo-Saxon perspective on trauma has mostly ignored this contribution to the trauma field, while practitioners in Latin countries (Southern Europe and Latin America) still use this approach in clinical practice. Contemporary French theory was examined in Chapter 3 in which the Freudian and Lacanian inspired trauma visions of Crocq, Lebigot and Barrois were discussed in the light of the trauma interpretation in DSM-IV and DSM-5 (APA, 1994, 2013).

Crocq (1999) describes the traumatic neurosis as an independent neurosis, i.e. as a structured and lasting neurotic disorder, including both specific symptoms (symptoms of repetition) and general neurotic symptoms such as anxiety and asthenia, and involving a typical organization (which in this case is a reorganization) of the personality which produces and perpetuates symptoms. Instead of focusing on the biological, behavioral and cognitive aspects of trauma, French theory focuses on the phenomenological variety and the underlying meaning of trauma symptoms. Clinically, it makes sense to focus on the adaptive ability of disaster survivors to reach a level of integration of the traumatic experience into one's psychic existence by transforming the 'traumatizing' experience into a narrative, accompanied by a catharsis of affects. Overcoming the experience of 'effroi de la mort' ('dread ot death') and reconnecting with the world of 'those who speak', followed by the cessation of the repetition syndrome, could allow disaster survivors to recover from their intrusion symptoms, persistent avoidances and negative alteration in cognitions and mood.

The question is whether or not researchers, aiming at quantitative descriptions of posttraumatic reactions, can bridge their scientific interpretation with the phenomenological approach of clinicians in non-English speaking countries.

9.2.2. Survivors of a dreadful and massive explosion experience vehement negative emotions which lead to a unique encounter with the 'real-of-death' and a 'repetition syndrome'.

At the phenomenological level, the overwhelming and vehement emotions at the moment of a massive explosion make it practically impossible to find a satisfactory narrative form for the experience of a survivor. The incapacity of the victim to perceive and personify the whole experience of a sudden enormous explosion cause a failure on the level of the memory: the event itself becomes a complex of disconnected visual, kinesthetic, physical and emotional sensation which may reappear on the surface with each sensation and/or emotion that recalls the original event (Van der Hart, Nijenhuis & Steele, 2006). This is described in Chapter 2 in which an individual survivor describes his struggle to survive.

Survivors of a disaster may get caught in a trap by the event and become a hostage of the many triggers that stay omnipresent due to extensive media coverage, the numerous contacts with caregivers and the specific medical condition – often related to long lasting burn injuries – in which survivors awake after the disastrous event.

The confrontation with the 'real-of-death' during the massive gas explosion, i.e. an encounter with the dread of death and the absolute 'nothingness' (Lebigot, 2002, 2004), penetrates deep into the psyche of the survivor (see Chapter 3). Every human being is aware of death as the inescapable end, but death itself does not have a representation in the psyche. The 'repetition syndrome' will appear, witnessing the 'effraction' of the psyche by traumatic images. Without symbolization, the 'new' reality about death, as a kind of special initiation to 'forbidden knowledge' (from a Lacanian point of view), will stay unintegrated. With the traumatic impact of the disaster experienced as incommunicable and inexpressible, survivors may stay blocked in their lived experience of abandonment and alienation.

Disaster survivors may also have to bear difficult feelings of culpability: the fact of surviving without having been able to help other victims. Next to these feelings, trauma survivors may give up their attempts for recognition which they expect from the surrounding community and the external world .

These phenomenological descriptions of psychological trauma may match with the current status of neurobiological knowledge about the traumatized brain. In our opinion, it makes sense to compare the accounts of the lived experiences of survivors with results from neurobiological research which describes the neurovegetative storm through which trauma victims go during intimate encounter with death. An extensive review of neurobiological trauma research is beyond the scope of this thesis, but the 'window of tolerance'-model (Ogden & Minton, 2000; Siegel, 1999) explains how victims of trauma may experience inflexible, rigid and chaotic states, during moments of extreme and maladaptive arousal. These states create the failure of the integrative system and might well correspond to the hermeneutically described effraction of the psyche (*effraction traumatique*), leading to a repetition syndrome.

9.2.3. In technological disasters, both during and in the aftermath, the traumatizing potential of the lived experiences of firefighters are based on event-specific and life-threatening characteristics while the experiences of emergency medical services (EMS) personnel are more oriented toward human suffering.

Chapter 4 provided the answer to a series of questions about the difference in the lived experiences of firefighters and EMS personnel or in-hospital staff. The phenomenological analysis of lived experiences of fire and emergency medical services personnel during their interventions in the context of the Ghislenghien disaster confirmed that EMS personnel are more directly involved with the injured victims.

The emotional impact of the disaster is related to the unusual number of victims and the massive human suffering. Firefighters involved in the Ghislenghien disaster had to cope with the event-specific and life-threatening characteristics of the explosion, i.e. respectively the sensory aspects of the explosion (e.g., heat, smoke, fire, smell, images of devastation) and the danger of additional explosions or secondary fires. It is understandable that in technological disasters, the direct contact with death and perceived life-threat contributes to the development of posttraumatic stress symptoms and emotional disturbances. Previous research (Laposa & Alden, 2003) confirmed that firefighters' work-related experiences are physically more dangerous and involve greater threat to their lives. The exposure with life-threat and death was imminent for the firefighters who worked on the site of the Ghislenghien disaster since they were confronted with the sudden death and the gruesome images of their five colleagues, wearing their protective gear but almost reduced to dust. The horror, the powerlessness and the emotional shock created by this terrible loss can be related to the experience of dread ('effroi') as described by Lebigot (2004). The concept of 'effroi' represents the very first state of frozen fright, petrified fright after the traumatizing blaze; a state believed to precede the development of stress and anxiety according to French contemporary trauma specialists. Compared to the firefighters, EMS personnel were less overwhelmed by the sudden and dangerous event, but the screaming, pain and suffering of the victims, including the smell of burned flesh, appeared to be at the center of their experience.

These results support the conclusion that the potentially traumatizing impact of the Ghislenghien disaster on firefighters is related to the event-specific characteristics of the explosion i.e. imminent life-threat, while the impact among EMS personnel (and in-hospital staff) is more related to the exposure massive amount of human suffering. This leads to a better understanding of the differences in the impact of terrifyingevents on rescuers and caregivers with a different role.

9.2.4. *(1) The prevalence of PTSD symptoms in adult survivors of a technological disaster (between 5 – 14 months) develops as a function of peritraumatic dissociation, exposure to the disaster, symptoms severity after a first measure in time and dissatisfaction with social support. (2) Mental health disturbances are more prevalent in the population that is closely involved in a technological disaster, as a function of their exposure to the human suffering of disaster survivors, compared to the reference population.*

In the adult survivors of the Ghislenghien gas explosion, the frequency of PTSD symptoms required for a full PTSD diagnosis was found to be a function of peritraumatic dissociation, exposure to the disaster, symptoms severity after a first assessment in time and dissatisfaction with social support. The symptoms appeared to be stable over time, i.e. 6% after four months and 6.6% over 14 months (see Chapter 6). This is the result of the fact that the majority of these disaster victims appear to recover over time while some show a late onset of symptoms. One should keep in mind that most respondents show resilience. This finding is in accordance with previous research about the Enschede disaster (Smid, Van der Velden, Gersons, & Kleber, 2012; Van der Velden et al., 2006). In the Enschede disaster, most survivors reported high initial intrusion and avoidance reactions and experienced a reduction of these symptoms over time. The survivors with a late onset were more likely than other participants in this study to use mental health services several years after the disaster; severe disaster exposure and perceived lack of social support were associated with late-onset symptoms.

Degree of exposure as a base for classification of disaster victims

Concerning the degree of exposure, the amount of potential life threat turned out to be an important determinant of the severity of PTSD symptoms. The majority of the survivors who were injured and hospitalized presented a full symptom pattern required for PTSD at respectively 5 (T1) and 14 months (T2) after the disaster. Still one out of three respondents who were directly exposed to the victims, either at the moment of the explosion, or because they were close to direct victims, presented a full PTSD symptoms pattern at T2. The results from the Ghislenghien disaster study clearly show that it is the degree of exposure to life-threat and/or human suffering that is related to the severity of symptoms.

Furthermore, the Ghislenghien disaster study indicated that the degree of exposure is useful for categorizing the disaster victims. The classification of disaster victims in previous research is often based on ad hoc criteria – residents of the disaster area, victims identified through the medical chain, people who declared themselves as victims by contacting supporting services, etc. Since there always seems to exist a dose-response relationship in the victims of a disaster, a more rigorous way of classifying victims may offer a solution for the diverse definitions of 'victims' found in current literature.

In previous publications (De Soir, 2006; Rooze, M., et al., 2008), the so-called CRASH approach (*Calamiteiten en Rampen Aanpak bij Slachtoffers en Hulpverleners – Calamity and Disaster Approach for Victims and Caregivers*) has been described as a model used for the triage of victims. The basis of this model is a 3 x 3 psychosocial matrix in which we find respectively in the rows and the columns: 1) *primary, secondary* and *tertiary victims,* belonging to one of these three categories depending on the type of impact i.e. degree of exposure suffered; and, 2) *primary, secondary* and *tertiary prevention,* depending on when the psychosocial support or assistance takes place. The concrete realisation of the framework for psychological (crisis) support consists, on the one hand, of a kind of psychological triage to sort out the different kinds of victims, dividing them in three different categories, and on the other hand, the selection of the appropriate support technique, at the right moment and carried out by the right people, thus trying to realise an optimal fit between victims and the type of support they get (De Soir, 2012).

Exposure to human suffering as a risk factor for mental health disturbances

From Chapter 7, it can be concluded that disaster survivors who have been confronted with an exposure to more human suffering, showed more mental health disturbances (such as somatization, depression, anxiety and sleeping disturbances) compared to more distant groups and the reference population. This is a clear indication for the existence of a dose-response relationship. Direct witnesses who had seen human damage had significantly higher prevalence rates regarding these mental health disturbances at T1 (5 months) and T2 (14 months). Significantly more often they suffered from somatization, depression, anxiety and sleeping disorders as compared to the reference population of the Province of Hainaut (in which the Ghislenghien gas explosion took place). Moreover, the impact of the disaster had not been resolved over time which is not consistent with previous studies (North, Kawasaki, Spitznagel & Hong, 2004; Shore & Tatum, 1986).

In line with older literature (Gatchel, Schaeffer & Baem, 1985; Ursano, Fullerton, Kao & Bhartiya, 1995), it can be concluded that the individual response to a disaster is related to event factors since the physical proximity to the disaster was a predominant criterion for the observed psychological impact (while the eventual effects of individual factors such as sex, age and educational level were controlled).

Peritraumatic dissociation and PTSD symptoms

Peritraumatic dissociation was a powerful predictor at T1 but showed no longer the same predictive value at T2. The symptoms severity at T1 appeared to be the largest predictor of a PTSD pattern at T2. Therefore, it is advisable to include peritraumatic dissociation in screening instruments intended to identify victims at risk for chronification of posttraumatic sequalae.

In the last decade researchers have struggled with the concept of 'peritraumatic dissociation', in particular after the introduction of the Acute Stress Disorder (ASD) diagnosis in DSM-IV, which required symptoms of dissociation such as derealisation, depersonalization, dissociative amnesia and a reduction of awareness (e.g. 'being in a daze'). The expectation was that PTSD could be predicted in an early stage. This expectation has been studied in a number of studies (Birmes et al, 2003; Fullerton et al., 2004; Meiser-Stedman, Yule, Smith, Dalgleish, & Glucksman , 2007; Nugent, Christopher, & Delahanty, 2006).

A critical feature of peritraumatic dissociation is the acute integrative failure, which sets the stage for the failure to synthesize and personify experience in the long run. However, in other studies (e.g., Holeva, Tarrier, & Wells, 2001) peritraumatic dissociation has not been found to be predictive of PTSD.

The controversy on the predictive role of peritraumatic dissociation in the development of PTSD may be related to the lack of knowledge on the relationship between peritraumatic dissociation and significant peritraumatic emotions, such as fear and loss of control (Gershuny, Cloitre, & Otto). Empirical studies conducted 15-20 years ago have suggested that dissociation is a response to overwhelming emotional and physiological arousal (Bernat, Ronfeldt, Calhoun, & Arias, 1998; Ehlers, Mayou, & Bryant, 1998; Simeon, Greenberg, Knutselka & Hollander, 2003), while other (often more recent) studies did not show this anymore or to a lesser extent after other variables were taken into account (e.g., Marshall & Schell, 2002; Marx & Sloan, 2005).

In our opinion, there is a lack of research in which the context of a specific event, such as a technological disaster, is taken into account when analysing the psychological responses to a potentially traumatizing event. The specific impact that a given event has on a survivor should be better analysed because since long peritraumatic dissociation has also been found to be related to greater perceived threat and greater externality in locus of control (Marmar et al., 1999), loss of control, helplessness and anxiety (Simeon, Greenberg, Knutselska, & Hollander, 2003), and hyperarousal and anxiety (Sterlini, & Bryant, 2002).

The account of the survivor of the massive gas explosion of the Ghislenghien disaster (see Chapter 2) illustrates that peritraumatic dissociation should be conceptualized as part of the overwhelming fright process, such as described by Gershuny et al. (2003), in which the cognitive appraisal of such a disaster as unmanageable may help elicit dissociation. Future research should focus on possible confounding variables in the relationship between peritraumatic dissociation and posttraumatic stress symptoms, as discussed by Van der Velden and Wittman (2008), and on the various types of contexts in which peritraumatic dissociation occurs. It is possible that the phenomenon of peritraumatic dissociation may vary according to the type of traumatization survivors experience.

Optimizing social support in the wake of disaster

Finally, the development of PTSD symptoms was positively related to dissatisfaction with the provided psychological and social support (see Chapter 6 and Chapter 8) but not with the number of people (potentially) providing support. This finding is in line with previous research (Drogendijk, Van der Velden, Gersons, & Kleber, 2011; Norris, Baker, Murphy, & Kaniasty, 2005). This connection between PTSD symptoms and dissatisfaction with the provided support indicates that the development of adequate instruments to assess the quality of care, on significant moments in the wake of disaster, is of crucial importance.

The structured help for disaster survivors should be tailored to the needs of victims belonging to the different target groups and organized at the community level on the basis of a scientifically sound instrument. It is important that the questioning of victim groups happens in a co-ordinated way, hand in hand with the multiple administrative forms that victims have to fill-in (see Chapter 5). Otherwise, questionnaires aimed at evaluating their post-disaster (mental) health and offering support might be seen as additional unnecessary red-tape in the wake of a disaster.

Next to a psychosocial supportive response towards the different victim groups, the provision of social support may benefit from modern technology and social media. A specially designed website and the intensive use of social media, with a purpose of dissemination of information and assistance, may consistently contribute to the recovery of the community of victims involved in a given disaster.

Public authorities should invest more in these aspects of support (also pre-disaster) instead of studying the occurring (mental) health problems retrospectively. Several chapters of this thesis (Chapter 5, 6 and 8) have indicated that dissatisfaction with insufficient social support over time will lead to problems in the post-disaster recovery of a community.

9.2.5. In child survivors of a technological disaster, the type of exposure, peritraumatic dissociation and received psychological help are related to the severity of PTSD symptoms.

The Ghislenghien disaster study showed that in the sample of responding children (between the ages of 8 and 14 years) 7.5% presented posttraumatic stress reactions at T1 (5 months) and 4.3% still showed these reactions at T2 (14 months). The PTSD prevalence in this sample seemed less stable over time than was the case in adult survivors.

A first risk factor associated with the severity of posttraumatic stress reactions in children was the type of exposure, which referred to the subjective disaster experience of the children. The perceived life threat, experienced by the children who were primary (or direct) victims, was the highest level of exposure. The secondary victims, who were direct witnesses of severe human suffering (e.g. burned, injured, and/or deceased victims), family members or close relatives of someone who was injured or died during the disaster, or direct witnesses of the explosion, were also affected, but to a lesser degree. Children only showed moderate to mild posttraumatic stress symptoms after being confronted with life threat and/or human suffering.

The second risk factor that came out of this study was peritraumatic dissociation during or immediately after the disaster. Peritraumatic dissociation was the most important factor in the development of posttraumatic stress symptoms at T1, which is in accordance with previous findings of Ozer et al. (2003).

Finally, the Ghislenghien study on children indicated that children should have received more help in the immediate aftermath of the disaster, according to the answers of their parents on the questions regarding the psychological support and the perceived benefits of that help. This finding indicates that more effort should be invested in the triage of children (and the parents) in need for help and that both the contents and the timing of this kind of support are important.

9.4. Methodological deliberations

In disaster research, it is often difficult to define the victims in the direct aftermath. The participants for the Ghislenghien study included all the residents living near the industrial site and the employees of the companies located on the disaster site, as well as family members living at the same address, including children from 8 to 14 years old. Family members of deceased and injured victims were also targeted. All of them were connected with the disaster through a geographical or professional proximity as well as connection through relatives.

The low response rates were a disappointment for the research team. The overall response rate at household level was 17% for the residents living within 5 km of the epicentre, 23% for the employees of companies and 25% for the people who identified themselves to an emergency service; the overall response rate only reached 18%. Fourteen months after the disaster, the response rate at household level was 56% (338 families); this is 9.8% of the initial study population who received a questionnaire five months after the disaster. These low response rates raise the question on the study representativeness. This low response rate is quite puzzling and leads to the important question about how to increase the voluntarily participation in post-disaster research. In my opinion, a more centralized approach in which different research teams work together, hand in hand with the psychosocial support structures, would assure a better response and increase the compliance of disaster survivors towards participating in scientific research.

Through several channels – personal contacts during victims' meetings, telephone contacts, written answers on open-ended questions, meetings with psychosocial caregivers, hospitals, etc. - the researchers understood that the majority of people who were directly involved but declined to participate did not want to be confronted with the disaster again by filling in questionnaires. It is acceptable that this problem has been more prominent among direct victims or witnesses of the disaster which could lead to an underestimation of the (mental) health outcomes of this study. Others probably did not participate in the study because they did not feel involved or concerned with the disaster. This is a common problem with postal surveys (CROSP/IHE, 1995). People might also have felt disappointed about the management of the disaster by the government. Furthermore, bureaucratic and political problems concerning the authorization and budgeting of this study, led to a considerable delay in the onset of this scientific investigation. It took about 4 months of negotiations with the authorities before the study was approved. The time interval between the gas explosion and the first point of measurement was 5 months. The researchers also dealt with considerable difficulties in the collection of the necessary addresses of the affected people and there was a lack of co-ordination between the different governmental departments (both local and federal), or organizations involved. Given the complex structure of the federal state which Belgium is, there is no culture of systematic disaster research and the key stakeholders of a fragile psychosocial disaster plan seemed to fear negative evaluation. Hopefully, these methodological considerations will lead to an improvement of future disaster research in Belgium with more accurate methods to contact people involved in a disaster.

Studies describing the impact of disaster on (mental) health always deal with considerable methodological weaknesses. Therefore, the comparison of results between different disaster studies needs to be carried out with caution. The problem is often related to the definition of the different exposure groups, as discussed in previous research (Brewin, Andrews, & Valentine, 2000; Raphaël & Wilson, 1993). Disaster studies may focus on the residents of the official disaster area (Yzermans, & Gersons, 2002), on survivors who had to be relocated after the disaster (Grievink, Van der Velden, Yzermans, Rooda & Stellato, 2006) or on victims who are identified through the medical chain and as a function of their medical condition (Dorn, Yzermans, Kerssens, Spreeuwenberg, & Van der Zee, 2006).

In the light of the above, next to the longitudinal design, the most original aspect of the Ghislenghien disaster study has probably been the classification of victims in 9 categories, as a function of their type/degree of exposure. The hypothesis that mental health problems of the defined groups would depend on the extent to which they experience life-threat and/or witnessed or suffered serious injury and human suffering, has been clearly confirmed. This approach also allowed to use the defined groups as conditions to be compared to each other.

9.3. Practical implications for peri- and post-disaster psychosocial support

The research which has been the basis of this thesis, confirms that a technological disaster produces sudden, unexpected, and "earth-shattering" effects on those affected. Disaster survivors often talk about how their lives have radically changed and testify how things are not the same anymore. Or how their inner sense of safety and their ability to count on the stability of their environment has been lost. Some also speak about feelings of powerlessness and having lost the structure of their daily lives. A disaster such as the explosion in Ghislenghien may affect the lives of many people for years, but still the default mode of the human being appears to be resilience and spontaneous recovery.

For those who will develop PTSD symptoms, the focus put on peritraumatic dissociation, degree of exposure, psychological help and social support, supports the idea that psychosocial intervention should be available, as a function of observable needs of survivors, from the first moments after the disaster. Therefore, paramedics, firefighters, and other rescue workers should be taught to provide the earliest psychological interventions to people in need, showing empathy in a victim-specific way and being sensitive to the immediate needs of victims.

Given the difficulties in the classification of disaster victims in retrospective disaster research, caregivers could start with psychological triage next to the current medical triage which is a part of the medical chain. Early psychological triage may appear to be useful in the psychosocial follow-up of the disaster.

Within the policy of medical assistance during times of disaster, psychological triage and early psychological support should be part of the planned activities during the acute phase. Immediate psychological support while the disaster still is in progress could be called 'peritraumatic support' and it can be provided by first responders from fire, rescue and police services. On-scene trauma support can take place by getting close to the victim, judging the situation of the victims, looking at the victims' emotional reactions, and assisting them by diverting their attention from pain and experienced life-threat.

While these supportive interventions, aiming at psycho-physical stabilization, seem adequate, future research should indicate more clearly what should or should not be done by first aid and rescue workers. Information of disaster victims and facilitating social support should be considered a psychosocial intervention since the answers on open-ended questions in the Ghislenghien study indicated that the perceived lack of support of the survivors had to do with missing the accurate, factual information on the incident itself and on where to get help.

In the wake of the Ghislenghien disaster, a lot of ad hoc supportive structures have been created, but most of them disappeared as quickly as they arose in the aftermath of the explosion. Surveys, scientific research, administration, legal inquiry, etc. should be organized and oriented towards the victims in a co-ordinated way from a centralized and governmental facility. Building different formalized groups of communicators at local, regional, and central levels helps gain knowledge of the need of each targeted group of victims.

9.6. Final conclusion

The first chapters in this thesis aimed to provide the 'real face of death' in technological disaster. We think that current literature lacks this kind of reporting on the subjective challenges of trauma survivors. The idiosyncratic truth of each survivor adds something unique to what is already known about psychological trauma. We wanted to provide an insight in the terrible reality of a technological disaster, starting from the individual and qualitative level, over a theoretical discussion, to the collective and quantitative level. We hope to have created a bridge between the point of view of the clinician and the more empirical position of the researcher.

Notwithstanding the methodological shortcomings, this study provided important knowledge about the long-term (mental) health consequences of a technological disaster. As in other countries, the concept of emergency and disaster planning in Belgium is in constant evolution. Since 1997, the Federal Public Service for Health has developed a Psychosocial Intervention Plan within the framework of a medical-sanitary aid service during collective emergencies.

The study on the disaster of Ghislenghien shows a series of possibilities for increasing the quality of both victims assistance and disaster research in Belgium. Its lessons should keep disaster managers humble. In order to provide the right psychosocial support during and in the aftermath of a disaster, a federal structure should be set up to deal with the co-ordination and the gathering of information related to the victims and their close relatives whereby the physical, mental and social health of the potentially affected population is systematically assessed and monitored. As long as such a structure will not exist, ad hoc disaster research will take time to be set up and may waste time for the recovery process of victims.

References

American Psychiatric Association. Diagnostic and Statistical Manual of Mental Disorders. 4th ed. Washington, DC: American Psychiatric Association Press; 1994

American Psychiatric Association. Diagnostic and Statistical Manual of Mental Disorders. 5th ed. Washington, DC: American Psychiatric Association Press; 2013.

Bernat, J., Ronfeldt, H., Calhoun, K., & Arias, I. (1998). Prevalence of traumatic events and peritraumatic predictors of posttraumatic stress symptoms in a nonclinical sample of college students. *Journal of Traumatic Stress, 11(4)*,645-664.

Birmes, P., Brunet, A., Carreras, D., Ducassé, JL, Charlet, JP, Lauque, D., Sztulman, H., Schmitt, L. (2003). The predictive power of peritraumatic dissociation and acute stress symptoms for posttraumatic stress symptoms: A three-month prospective study. *American Journal of Psychiatry, 160*,1337-1339. Doi: 10.1176/appi.ajp.160.7.1337.

Brewin, C., Andrews, B., & Valentine, J. (2000). Meta-analysis of risk factors for posttraumatic stress disorder in trauma-exposed adults. *Journal of Consulting and Clinical Psychology, 68(5)*,748-66.

Crocq, L. (1999). *Les traumatismes psychiques de guerre.* Paris: Odile Jacob.

CROSP/IHE-Ecole de Santé Publique (1995). *Enquête de santé par interview: Aspects méthodologiques, 47.* Bruxelles: Document de travail.

De Soir, E. (2006). *Een heel klein beetje oorlog... Omgaan met traumatische ervaringen.* Tielt: Lannoo.

De Soir, E. (2012). The management of emotionally disturbing interventions in fire and rescue services: psychological triage as a framework for acute support. In R. Hughes, C. Cooper, & A. Kindler (Eds.): *International Handbook of Workplace Trauma Support* (pp.368-383). New York: John Wiley & Sons.

Dorn, T., Yzermans, C., Kerssens, J., Spreeuwenberg, P.,& Van der Zee, K. (2006). Disaster and subsequent healthcare utilization. A longitudinal study among victims, their family, and control subjects. *Medical Care, 44(6)*,581-589.

Drogendijk, A., Van der Velden, P., Gersons, B., & Kleber, R. (2011). Lack of social support among ethnic minorities after a disaster: Comparative study. *British Journal of Psychiatry, 198*,317-322.

Ehlers, A., Mayou, R., & Bryant, B. (1998). Psychological predictors of chronic post-traumatic stress disorder after motor vehicle accidents. *Journal of Abnormal Psychology, 107*,508-519.

Gatchel, R., Schaeffer, M., & Baum, A. (1985). A psychophysiological field study of stress at Three Mile Island. *Psychophysiology, 22(2)*,175-181.

Gershuny, B., Cloitre, M., & Otto, M. (2003). Peritraumatic dissociation and PTSD severity: Do event related fears about death and control mediate their relation? *Behaviour Research and Therapy, 41*,157-166.

Grievink, L., Van der Velden, P., Yzermans, J., Rooda, J., & Stellato, R. (2006). The importance of estimating selection bias on prevalence estimates shortly after disaster. *Annals of Epidemiology, 16(10)*,782-286.

Holeva, V., Tarrier, N., & Wells, A. (2001). Prevalence and predictors of acute stress disorder and PTSD following road traffic accidents: Thought control strategies and social support. *Behavior Therapy,32*,65-83.

Laposa, J., & Alden, L. (2003). Posttraumatic stress disorder in the emergency room: exploration of a cognitive model. *Behaviour Research & Therapy, 41(1)*,49-65.

Lebigot, F. (2002). Le traumatisme psychique. Le regarder en face ou s'en protéger. *Revue francophone du stress et du trauma, 2(3)*,139-146.

Lebigot, F. (2004). Le traumatisme psychique. *Revue francophone du stress et du trauma, 4(1)*,5-11.

Marmar, C., Weiss, D., Metzler, T., Delucchi, K., Best, S., & Wentwort, K. (1999). Longitudinal course and predictors of continuing distress following critical incident exposure in emergency personnel. *Journal of Nervous and Mental Disease,187*,15-22.

Marshall, G., & Schell, T. (2002). Reappraising the link between peritraumatic dissociation and PTSD symptom severity: Evidence from a longitudinal study of community violence survivors. *Journal of Abnormal Psychology,111*,626-636.

Marx, B., & Sloan, D. (2005). Peritraumatic dissociation and experiential avoidance as predictors of posttraumatic stress symptomatology. *Behaviour Research and Therapy,43*,569-583.

Meiser-Stedman, R., Smith, P., Yule, W., Dalgeish,T., & Glucksman, E. (2007). Diagnonostic, demographic, memory quality, and cognitive variables associated with acute stress disorder in children and adolescents. *Journal of Abnormal Psychology, 116(1)*,65-79. doi: 10.1037/0021-843X.116.1.65.

Norris, F., Baker, C., Murphy, A., & Kaniasty, K. (2005). Social support mobilization and deterioration after Mexico's 1999 flood: Effects of context, gender, and time. *American Journal of Community Psychology,36*,15-28.

North, C., Kawasaki, A., Spitznagel, E., & Hong, B. (2004). The course of PTSD, major depression, substance abuse, and somatization after a natural disaster. *Journal of Nervous and Mental Disease, 192(12)*,823-829.

Nugent, N., Christopher, N., & Delahanty, D., (2006). Emergency medical service and in-hospital vital signs as predictors of subsequent PTSD symptom severity in pediatric injury patients. *Journal of Child Psychology and Psychiatry, 47(9)*,919-926.

Ogden, P., & Minton, K. (2000). Sensorimotor psychotherapy: One method for processing trauma. *Traumatology, 6.*

Ozer, E., Best, S., Lipsey, T., & Weiss, D. (2003). Predictors of posttraumatic stress disorder and symptoms in adults: A meta-analysis. *Psychological Bulletin, 129*,52-73.

Raphael, B., & Wilson, J. (1993). Theoretical and intervention consideration in working with victims of disaster. In: J. Wilson & B. Raphael (Eds.), *The international handbook of traumatic stress syndromes.* New York: Plenum.

Rooze, M., Netten, J., De Ruyter, A., De Vries, M., Helsloot, I., De Soir, E., Selwood, P., a Schenk, H. **(2008).** Prioritizing Care During the Acute Phase: The Prominent Role of Basic Psychosocial Life Support. *Prehospital and Disaster Medicine.* July-August 2008.

Shore, J., Tatum, E., & Vollmer, W. (1986). Psychiatric reactions to disaster: the Mount St Helens experience. *American Journal of Psychiatry, 143(5)*,590-595.

Siegel, D. (1999). *The developing mind: toward a neurobiology of interpersonal experience.* New York: Guilford.

Simeon, D., Greenberg, J., Knutselka, M., Schmeidler, L., & Hollander, E. (2003). Peritraumatic reactions associated with the World Trade Center disaster. *American Journal of Psychiatry, 160*, 1702-1705.

Smid, G., Van der Velden, P., Gersons, B., & Kleber, R. (2012). Late-onset posttraumatic stress disorder following a disaster: A longitudinal study. *Psychological Trauma: Theory, Research, Practice, and Policy, 4(3)*,312.

Sterlini, G., & Bryant, R. (2002). Hyperarousal and dissociation: A study of novice skydivers. *Behavioral Research and Therapy, 40*,431-437.

Van der Hart, O., Nijenhuis, E., & Steele, K. (2006). *The haunted self: Structural dissociation and the treatment of chronic traumatization.* New York: Norton.

Ursano, R., Fullerton, C., Kao, T., & Bhartiya, V. (1995). Longitudinal assessment of posttraumatic stress disorder and depression after exposure to traumatic death. *Journal of Nervous and Mental Disease, 183(1)*,36-42.

Van der Velden, P., Van der, Grievink, L., Kleber, R., Drogendijk, A., Roskam, A., Marcelissen, F., Olff, M., Meewisse, M., & Gersons, B. (2006). Post-disaster mental health problems and the utilization of mental health services: A four year longitudinal comparative study. *Administration and Policy in Mental Health and Mental Health Services Research, 33,* 279-288.

Van der Velden, P., & Witman, L. (2008). The independent predictive value of peritraumatic dissociation for PTSD symptomatolgy after type 1 trauma: A systematic review of prospective studies. *Clinical Psychology Review, 28(6)*,1009-1020.

Yzermans, J., & Gersons, B. (2002). The chaotic aftermath of an airplane crash in Amsterdam. In: J. Havenaar, J. Cwikel, & E. Bromet (Eds.). *Toxic turmoil: psychological and societal consequences of ecological disasters.* New York: Kluwer Academic/Plenum Publishers.

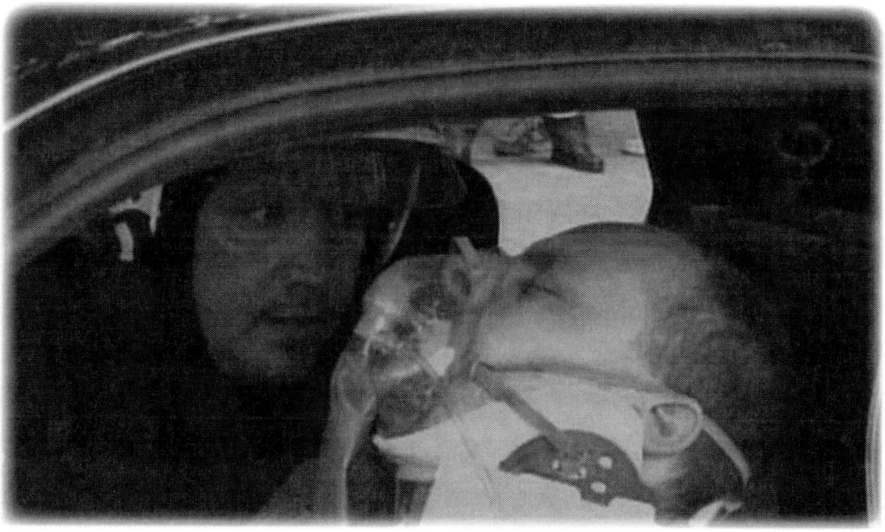

EPILOGUE

Is the worst really over after the crash?!

One day, maybe sooner, maybe later, someone you love will be in an emergency situation. If the crisis hits and you are in the place of the first responder, after you have called in for professional help, and, if appropriate, propped up the victim's head or feet, what do you do? What do you say? 'Hang in there, Joe,' or 'don't die on me, damn it,' may be heartfelt but is not exactly helpful. "But there ARE words, and ways of saying them, that are not only helpful, but can turn the situation around, can positively affect Joe's heart rate, temperature, breathing, in fact his cardiovascular, limbic, endocrine, circulatory and respiratory systems."

From: 'The Worst is Over: What to Say When Every Moment Counts – Verbal First Aid to Calm, Relieve Pain, Promote Healing, and Save Lives' – Judith ACOSTA & Judith Simon PRAGER (2002).

Is the worst really over after the crash?!

Research (Acosta & Prager, 2002) indicates that saying the right words at the right time in the right way can change the outcome of critical care and set the course of recovery for people even before they arrive in the emergency room. People in crisis are in an altered state of consciousness while experiencing an emotional crisis or medical emergency, which is precisely why what you say is as important as what you do. And the right words said in a frightening situation may help shift the memory from one of horror and fear into a memory of rescue and perhaps even courage. Because we are a mind-body, physically and emotionally words can harm or words can heal.

The principle involved in using language to connect with the body and spirit for healing applies anywhere, everywhere, to all of us whenever sensitivity, understanding, kindness, and conviction are required. Many medical professionals and first responders around the world are using it to calm, relieve pain, promote healing, and save lives. And all of us, because we all have words and the ability to use our empathy and kindness, are ultimately equipped to help start the healing and change the outcome for the better. Will you know what to say when every moment counts?

This epilogue aims to reflect on the possibilities of advanced psychological support through psycho-physiological stabilization in cases where you can be that invaluable resource.

My experiences in the field, as a voluntary paramedic and firefighter, indicated that victims of motor vehicle accidents (MVAs) often develop psychological problems due to contextual aspects, such as the life-threatening character, the disorientation, the anxiety, the pain and the loss of control, inherent in severe traffic accidents. I visited a lot of survivors of MVAs in hospital, trying to document their lived-experiences through a semi-structured interview. I would bring them sympathy and attention from my colleagues, the firefighters and paramedics. They wished for news of the victims' situation and to hear about the enormous challenges associated with the extrication procedures; survivors often felt reduced to 'human material' in the process of being uncovered from a car wreckage. Some of these survivors came into my therapy room, months or even years after surviving an accident in which they were critically injured.

In these years, I developed the 'fire rescue squirrel method' on a rather intuitive basis. The 'ResQSquirrel' is the individual firefighter/paramedic, who is certified in emergency crisis response and advanced psychological support, and who provides emergency medical and psychological stabilisation for (critically injured) victims. Most commonly, this involves victims entrapped in a vehicle after an accident. The name 'squirrel' metaphorically refers to the characteristics of these first responders: quick, swift, intelligent, action-oriented and gentle. The ResQSquirel aims at a positive impact on the victims' vital parameters (heart rate, respiratory frequency and blood pressure) and the psychological experience both during and immediately after the rescue operation (De Soir & Goffings, 2009, 2012).

For the first assessment of the utility of the ResQSquirrel-method, I developed a specific training package for first responders from emergency medical and fire services, based on (hypnotic) techniques for arousal, pain and anxiety reduction. With this training, these first responders were taught how to anticipate the direct needs of MVA victims in order to decrease peritraumatic discomfort and reduce fear, pain and disorientation with the aim of minimizing the development of subsequent psychological posttraumatic disturbances.

Post-accident interviews with MVA survivors made me enthusiastic about our attempts to assist critically injured victims: most survivors reported a strong feeling of safety and even of being in peace. The next statements (cf. infra) used by the casualties in the interview show what aspects of the support they named as important.

> *"Just the fact that someone was near me and knew what the others were doing, that he was talking to me; calmed me down. He had a very soothing voice and said to me 'listen to my voice and imagine both of us talking through a telephone'. (...) He gently holds my hand, explaining each step in the extrication procedure. I was prepared to hear each new sound, even the intense and extreme ones, which were labeled as 'that's OK; it shows that a whole crew is working to get you out of here'* (Janet, 26 years). "

The fact that the rescue squirrels are always both trained fire fighters and paramedics, provides the advantage that they know all the technical aspects of rescuing and freeing a victim from a car. Thereby reducing the effect of surprise and powerlessness for the victim appears to be essential.

An extreme sense of powerlessness was reported by one of the victims. After reviewing the interview, it appeared that this feeling was overwhelming in the first minutes after the crash when the rescue squirrel was not yet operational and that it disappeared after the first moments of contact.

> *"The powerlessness was very frustrating, it seemed as if I stayed in that wreckage, alone and completely abandoned for hours. But once a firefighter entered my [sic] accidented car and started with his soothing techniques, I really felt in good hands. On the most difficult moments of my liberation, he always asked for my approval or hand sign, 'thumb up', before giving his colleagues the permission to proceed. It gave me a strange sense of control over a situation which really was totally out of control* (Danny, 47 years)"

Assisting the victims with their breathing, explaining the sequence of the extrication procedure, providing them a subjective sense of control during the extrication procedure and involving them in the extrication process, appears to be very soothing. The rescue squirrel reaches positive results in a rather easy way, saying things like: *"Just say yes when you want us to continue..., are you okay? Are you ready, you will hear quite some noise but I'm right beside you and I will stay here the whole time, there is nothing to worry about, you are in good hands, me and my colleagues, we know what to do in these situations. Trust your body, it also knows exactly how to adapt to this special situation. It knows how to keep in control* (...)".

Easy things, such as protecting victims with a transparent hard plastic board in order to avoid the debris from hurting them, and keeping them oriented, provides a feeling of safety and an opportunity to stay aware of the total environment. Loss of control can induce a more explicit perceived threat since MVA victims do not understand the technical manoeuvres in the correct sense.

In each interview with an accident survivor, I asked about the perception of time during the accident. They mentioned a shortened perception of time. Two participants in our pilot study have encountered both a shortened and a prolonged perception of time. After reviewing these specific interviews, it could be concluded that the victims reported a prolonged perception of time before there was any contact with the fire squirrel.

These preliminary experiences have been used to convince the leadership of fire, rescue and emergency medical services that training first responders to use simple techniques for psychological stabilization which can make a world of difference for critically injured MVA victims who have to be extricated.

In order to validate these anecdotic findings, there is a need to conduct comprehensive research (ideally, randomized, controlled and international). The research of this thesis has been a cornerstone for these further developments; developments which are significant steps towards helping victims to gain mental and physical control over uncontrollable events.

References

Acosta, J. &, Prager, J.S. (2002). *The Worst is Over: What to Say When Every Moment Counts – Verbal First Aid to Calm, Relieve Pain, Promote Healing, and Save Lives.* San Diego, CAiJodere Group.

De Soir, E. & Goffings, K. (2009). Psychological stabilization for MVA victims. *Crisis Response.* 5(4),40-41.

De Soir, E. & Goffings, K. (2012). Directe opvang van verkeersslachtoffers. In E. Vermetten, R.J. Kleber, & O. van der Hart (Eds.), *Handboek posttraumatische stressstoornissen* (pp. 321-335). Utrecht: De Tijdstroom.

Samenvatting in het Nederlands

Dit proefschrift behandelt de geestelijke gezondheidsproblemen in de nasleep van de gasontploffing in Gellingen (Ghislenghien, 30 juli 2004) in België waarbij vijf brandweerlui, een politieman en achttien andere slachtoffers stierven. Deze ramp maakte ook 132 slachtoffers van wie velen met (ernstige) brandwonden.

Het eerste hoofdstuk schetst de onderzoeksvragen van deze thesis: een algemene doelstelling is het bestuderen en het verduidelijken van de weerslag van een technologische ramp, zowel op empirisch als op fenomenologisch vlak. Deze thesis bestaat daarom uit een fenomenologisch deel en een empirisch deel.

Vanuit een fenomenologisch oogpunt werd de beleving van een overlevende van deze gasontploffing gebruikt als achtergrond om de conceptuele verschillen te bespreken tussen de Angelsaksische traumatheorieën enerzijds en de meer klassieke psychodynamische theorieën die nog steeds populair zijn in de Latijnse landen, anderzijds.

Vanuit een empirisch standpunt waren de onderzoeksvragen gericht op de prevalentie van symptomen van posttraumatische stress symptomen bij volwassenen en kinderen die betrokken waren bij deze gasramp, zowel bij de familieleden van de overlevenden als bij de familieleden van de overledenen. Andere onderzoeksvragen hadden betrekking op de risicofactoren met betrekking tot de ontwikkeling van posttraumatische stress-symptomen en andere geestelijke gezondheidsproblemen in de verschillende groepen slachtoffers.

In de marge van deze onderzoeksvragen, was het ook de bedoeling om meer inzicht te verkrijgen in de uitdagingen met betrekking tot de aard van de psychologische hulp en het tijdstip waarop deze moet worden verschaft in de nasleep van een technologische ramp.

Het fenomenologische deel is opgebouwd uit drie hoofdstukken die contextuele en hermeneutische informatie bevatten omtrent de psychische problematiek veroorzaakt door een grootschalige explosie. Het empirische deel bestaat uit vier hoofdstukken die een kwantitatieve benadering bevatten van de trauma-gerelateerde geestelijke gezondheidsproblemen bij volwassenen en kinderen in de nasleep van een technologische ramp.

Hoofdstuk 2 is het eerste hoofdstuk van het fenomenologische deel en beschrijft de ervaringen van een overlevende van de gasontploffing van Ghislenghien aan de hand van zijn persoonlijke getuigenis over de progressieve traumatisering tijdens zijn overlevingsweg. Zijn beleving en ervaringen worden geduid tegen de achtergrond van zowel de Angelsaksische empirische traumamodellen als vanuit de psychodynamische theorieën die werden vormgegeven door invloedrijke hedendaagse Franse auteurs.

Een derde hoofdstuk onderzoekt de conceptualisatie van het psychologisch trauma in de hedendaagse Franse theorie en de klinische status die deze benadering nog steeds behoudt buiten de Angelsaksische psychotraumatologie. Het bespreekt ook de potentiële bijdrage van hedendaagse Franse auteurs aan de huidige kennis over psychotraumatologie.

Hoofdstuk 4 geeft een fenomenologische analyse van de ervaringen van brandweer- en urgentiepersoneel tijdens en onmiddellijk na de gasontploffing van Ghislenghien. Deze belevingen van brandweer- en urgentiepersoneel worden beschreven in termen van centrale kernmerken van potentieel traumatiserende gebeurtenissen en de meest schokkende aspecten van de ervaringen met betrekking tot deze ramp.

Hoofdstuk 5 is het eerste hoofdstuk van het empirische deel en baseert zich op een longitudinale studie die werd uitgevoerd om de weerslag van deze ramp op de fysieke, de mentale en de sociale gezondheid te bepalen en de prevalentie van posttraumatische stress-symptomen in de betrokken populatie in kaart te brengen. Deze studie omvat volwassenen (≥ 15 jaar) en kinderen (8-14 jaar) die het risico liepen om negatieve gezondheidseffecten te ontwikkelen in de nasleep van deze ramp (5 en 14 maanden erna). Dit hoofdstuk rapporteert de sterktes en zwaktes van de gebruikte methode en het mogelijke effect daarvan op het responspercentage. De ondervonden problemen en uitdagingen die opgetreden zijn tijdens de uitvoering van deze studie worden besproken tegen de achtergrond van de kenmerken van de ramp.

Hoofdstuk 6 analyseert de risicofactoren voor de ontwikkeling van posttraumatische stress-symptomen in de verschillende groepen slachtoffers, 5 en 14 maanden (respectievelijk T1 en T2) na de ramp, die betrokken waren in deze ramp. Hiërarchische regressie werd aangewend om de significante risicofactoren te bepalen alsook de mate van variantie die ze verklaren in het model vast te leggen. De symptomen van de verschillende categorieën getroffenen tonen duidelijk de invloed van het type blootstelling, peritraumatische dissociatie en de onvrede met de ervaren sociale steun. Op T1 was peritraumatische dissociatie de belangrijkste risicofactor met betrekking tot de ontwikkeling van symptomen van posttraumatische stress, maar op T2 waren posttraumatische stress-symptomen (PTS) op T1 de belangrijkste voorspeller van PTS symptomen op T2. Onvrede met sociale steun hing samen met de ontwikkeling van PTS symptomen op T1 en persistentie van deze symptomen op T2. Op T1 vertoonden 6% van de respondenten voldoende symptomen om een diagnose van Posttraumatische Stress Stoornis (PTSS) te krijgen. Op T2 vertoonden 6,6% voldoende symptomen om de diagnose van PTSS te krijgen.

Hoofdstuk 7 onderzoekt het risico om vier types geestelijke gezondheidsproblemen (*mental health disturbances*, MHD) te ontwikkelen als gevolg van blootstelling aan de verschillende aspecten van deze ramp en dit in vergelijking met gegevens verkregen uit eerder onderzoek bij een referentiepopulatie van dezelfde provincie. De "Symptom Checklist-90-Revised" (SCL-90-R) werd gebruikt om een prevalentieratio met betrekking tot somatisatie-, depressie-, angst- en slaapstoornissen te berekenen voor drie categoriën slachtoffers: directe getuigen van de menselijke schade (*seen human damage* of SHD), directe getuigen die geen menselijke schade gezien hebben (*not seen human damage* of NSHD) en indirecte getuigen (*indirect witnesses* of IW). De relatieve risico's werden ingeschat aan de hand van modellen voor logistische regressie. De prevalentieratio's waren voor de vier MHD's veel hoger bij respondenten uit de categorie 'directe getuigen van menselijke schade' dan voor de andere categorieën, zowel op T1 als op T2. Bovendien vertoonden de respondenten uit de twee andere categorie geen verhoogd risico op de ontwikkeling van één van de vier vormen van geestelijke gezondheidsproblemen vergeleken met de referentiepopulatie van de Provincie Henegouwen. Respondenten uit de categorie directe getuigen van menselijke schade vertoonden op T1 en T2 ruim vijf keer zoveel risico op de ontwikkeling van een somatisatie-stoornis, vier keer meer risico op de ontwikkeling van depressieve- en slaapstoornissen en vijf à zes maal keer risico op de ontwikkeling van een angststoornis. Tenslotte vertoonden ze op T1 en T2 respectievelijk 13 en 17 keer meer kans op het ontwikkelen van de vier genoemde geestelijke gezondheidsstoornissen samen.

Hoofdstuk 8 behandelt de risicofactoren met betrekking tot de ontwikkeling van posttraumatische stressreacties bij kinderen na de technologische ramp in Gellingen. De reacties van de betrokken kinderen werden gemeten na 5 maanden (T1, N-128) en 14 maanden (T2, N-69). Op T1 en T2 vertoonden respectievelijk 7% en 4% van de kinderen ernstige posttraumatische reacties. Bij de kinderen die op T1 posttraumatische reacties vertoonden, herstelde 60% op T2. De risicofactoren verbonden met posttraumatische stress reacties waren de mate van blootstelling aan de ramp, peritraumatische dissociatie en de mate van ontevredenheid met de gekregen psychologische hulp.

Hoofdstuk 9 verschaft de discussie en de conclusies van deze thesis tegen de achtergrond van de specifieke uitdagingen waarmee een gemeenschap geconfronteerd wordt na confrontatie met een technologische ramp. De resultaten van dit onderzoek bevestigen een aantal bekende risicofactoren met betrekking tot. de ontwikkeling en/of chronificatie van posttraumatische stress-symptomen, zoals peritraumatische dissociatie, type/maat van blootstelling aan de ramp en het ontbreken van sociale steun.

De epiloog aan het einde van deze thesis schetst de mogelijkheden voor een innovatieve en grensverleggende aanpak met betrekking tot de psychologische stabilisatie die een wereld van verschil zou kunnen betekenen voor de overlevenden van ongevallen en rampen voor wie het ergste niet over is na de ontploffing.

Summary in English

This thesis describes the mental health disturbances in the wake of the Ghislenghien gas explosion (July 30th 2004) which instantly killed five firefighters, one police officer and 18 other people. Moreover, 132 people were wounded among which many suffered severe burn injuries.

A first chapter provides the research questions of this thesis: in general it aims at studying and clarifying the impact of a technological disaster, both phenomenologically and empirically.

On the phenomenological side, the lived experiences of a disaster survivor are used to set the stage for a discussion on the conceptual differences between mainstream trauma theories (currently used by English-speaking researchers and practitioners) and the more classical psychodynamic theories (still being the basis for clinical practice in Latin countries).

On the empirical side, this study focuses on research questions regarding the prevalence of posttraumatic stress symptoms in adult and child survivors of a massive gas explosion, in their family members as well in family members of deceased victims. Other research questions focused on the risk factors regarding the development of posttraumatic stress symptoms and other mental health disturbances in different survivor groups.

In the margin of these research questions, the aim was also to better understand the challenges of the type and timing of psychological help in the wake of technological disaster.

Therefore, this manuscript consists of a phenomenological part, an empirical part and an epilogue.

The phenomenological part encompasses three chapters which provide contextual and hermeneutical information on the trauma inflicted by a massive explosion. The empirical part is composed of four chapters providing a quantitative approach of trauma-related mental health disturbances in adults and children after a technological disaster.

Chapter 2 is the first chapter of the phenomenological part and describes the experiences of a survivor (S.D.) of the Ghislenghien disaster by means of his testimony regarding the *progressive traumatisation* of his survival. The lived experiences of S.D. are explained relying on both empirical (Anglo-Saxon) trauma models and the psychodynamic theories developed by influential contemporary French authors.

The third chapter examines the conceptualization of psychological trauma in contemporary French theory and the clinical importance this approach still holds beyond the community of Anglo-Saxon trauma clinicians. It investigates the potential contribution from contemporary French authors to current knowledge on psychotraumatology.

Chapter 4 explores the experiences in fire and emergency services personnel during and immediately after the Ghislenghien gas explosion using a phenomenological approach. It describes the lived experiences of fire rescue and emergency services personnel engaged in the rescue operations during and immediately after a technological disaster in terms of central characteristics of potentially traumatizing events and the most shocking aspects of the experiences related to this disaster.

Chapter 5 is the first chapter of the empirical part and sets the base for a longitudinal study conducted in order to assess the impact of the disaster on physical, mental and social health, and to evaluate the prevalence of posttraumatic stress symptoms in the affected population. The study included adults (≥ 15 years) and children (8-14 years) at risk of developing adverse health effects related to the disaster at 5 months and 14 months after the disaster. This chapter reports on the strengths and weaknesses of the methodology employed and its influence on response rate. Difficulties and challenges encountered during the design of the study are discussed in relation to the characteristics related to this disaster.

Chapter 6 investigates the risk factors for the development of posttraumatic stress symptoms in the different survivor groups involved in the Ghislenghien disaster at 5 months (T1) and 14 months (T2). Hierarchical regression was used to determine the significant risk factors and to assess their proportion in variance accounted for. The symptoms of the different victim categories clearly indicate the influence of the degree of exposure, peritraumatic dissociation and dissatisfaction with social support on the development of posttraumatic stress symptoms. Peritraumatic dissociation appeared to be the most important risk factor of the development of posttraumatic stress symptoms at T1. But at T2, posttraumatic stress symptoms at T1 had become the most important predictor. Dissatisfaction with social support was positively linked to development of posttraumatic stress symptoms at T1 and to the maintenance of these symptoms at T2. At T1 6,0% of the respondents showed sufficient symptoms to meet all criteria for a full PTSD. At T2, 6,6% still suffered from enough posttraumatic stress symptoms to meet all PTSD criteria.

Chapter 7 explores the risk for the development of four types of mental health disturbances (MHD) due to exposure to different aspects of this disaster in comparison with data obtained from previous health surveys among the population of the same province. The "Symptom Checklist-90-Revised" (SCL-90-R) was used in order to compute actual prevalence rates of somatization-, depression-, anxiety- and sleeping disturbances for three defined exposure categories: direct witnesses who have seen human damage (SHD), direct witnesses who have not seen human damage (NSHD) and indirect witnesses (IW). Relative risks were estimated using logistic regression models. Prevalence rates of the four MHD were much higher for the SHD than for the other exposure groups, at T1 and T2. Moreover, NSHD and IW had no increased risk to develop one of the 4 types of MHD compared to the reference population. The SHD had at T1 and T2 good 5-times a higher risk for somatization, about 4-times for depression and sleeping disorders, and 5- to 6-times for anxiety disorders respectively. Further, they suffered 13 times, respectively 17 times more from all mental disorders together.

Chapter 8 focuses on the risk factors for the development of posttraumatic stress reactions in children after the technological disaster in Ghislenghien. Children were also assessed at five months (T1, N=128) and at fourteen months (T2; N=69). At T1 and T2 respectively, 7 % and 4 % of the responding children showed severe posttraumatic stress reactions. Of those who showed posttraumatic stress reactions at T1, 60% recovered from these symptoms at T2. Risk factors related to posttraumatic stress reactions were type of exposure to the disaster, peritraumatic dissociation, and dissatisfaction with the received psychological help.

Chapter 9 offers the discussion and conclusions of this thesis in the light of the specific challenges which may arise when a technological disaster suddenly strikes a community. The results confirm some of the well-known risk factors for the development and/or chronification of posttraumatic stress symptoms, such as peritraumatic dissociation, type/degree of exposure and lack of social support.

The epilogue describes the innovative and ground-breaking techniques of psychological stabilization that can make a world of difference for survivors of accidents since this research indicates that the worst might not be over after the blaze.

Résumé en Français

Cette thèse traite des troubles de santé mentale consécutifs à l'explosion d'un gazoduc survenue le 30 juillet 2004 à Ghislenghien, dans laquelle cinq pompiers, un policier et dix-huit autres victimes ont perdu la vie. La catastrophe a aussi causé 132 blessés, parmi lesquels de nombreux avec de graves brûlures.

Un premier chapitre pose les questions de recherche de cette thèse. De manière générale, l'objectif en est d'étudier et, par là même, de clarifier l'impact d'une catastrophe technologique, à la fois de manière phénoménologique et empirique.

Pour l'aspect phénomé-nologique, le vécu d'un survivant de la catastrophe sert de point de départ à une discussion portant sur les différences conceptuelles existant entre les théories empiriques (surtout utilisées par les praticiens anglo-saxons du trauma) et les théories psycho-dynamiques du trauma, plus classiques, qui servent de référence à la pratique clinique dans les pays latins.

Dans la partie empirique, l'accent est mis sur des questions de recherche portant sur la prévalence des symptômes de stress post-traumatique parmi les survivants de l'explosion, tant parmi les victimes adultes que chez les enfants, leurs proches, et les proches des victimes décédées. D'autres questions de recherche visent à étudier les facteurs de risque concernant le développement de symptômes post-traumatiques et d'autres troubles de santé mentale consé-cutifs à cette catastrophe. En filigrane à ces questions de recherche, le but est également de mieux comprendre l'importance et l'enjeu du type et de la programmation dans le temps des actions d'aide psychologique nécessaires après une catastrophe technologique.

La partie phénoménologique est constituée de trois chapitres qui fournissent l'apport contextuel et herméneutique du traumatisme psychique causé par une explosion massive. La partie empirique est quant à elle composée de quatre chapitres où sont présentés les résultats d'une étude quantitative des troubles de santé mentale liés au trauma parmi les personnes victimes de cette catastrophe.

Le chapitre 2 fait office de porte d'entrée à la partie phénoménologique décrivant le vécu d'un survivant (S.D.) de la catastrophe de Ghislenghien, portant témoignage d'une *traumatisation progressive* dans son parcours de survie. Les expériences vécues par S.D. sont expliquées tant en ayant recours aux modèles empiriques anglo-saxons du trauma qu'en se basant sur les théories psycho-dynamiques soutenues par les auteurs français contemporains qui font aujourd'hui autorité dans leur domaine.

Le troisième chapitre porte sur la conceptualisation du traumatisme psychique dans la théorie française contemporaine et traite de l'influence clinique de cette approche en-dehors de la communauté des cliniciens anglo-saxons du trauma. Aussi, ce chapitre se penche sur la contri-bution potentielle des auteurs français contemporains au domaine de la psycho-traumatologie.

Le chapitre 4 étudie de façon phénoménologique les expériences vécues par le personnel des services d'incendie et par les acteurs de l'aide médicale urgente durant et dans les suites immédiates de l'explosion survenue dans le zoning industriel de Ghislenghien. Aussi, ce chapitre fournit-il une description des ces expériences vécues en lien aux opérations de secours à la fois en termes de caractéristiques centrales d'événements potentiellement traumatisants et des aspects les plus choquants des expériences liées à cette catastrophe.

Le chapitre 5, premier de la partie empirique, présente l'étude longitudinale conduite afin d'envisager l'impact de la catastrophe sur la santé physique, mentale et sociale de la population atteinte, et d'en évaluer la prévalence des symptômes de stress post-traumatique. Cette étude porte sur les survivants adultes (≥ 15 ans) et les enfants (8-14 ans) présentant le risque de développer des problèmes de santé en lien à la catastrophe, et ce respectivement 5 et 14 mois après la catastrophe. Ce chapitre rapporte les points forts de cette étude ainsi que les faiblesses méthodologiques et leur influence sur le taux de réponse. Les difficultés rencon-trées et les défis surmontés dans le cadre de cette étude sont discutés à la lumière des caractéristiques propres à cette catastrophe.

Le chapitre 6 analyse les facteurs de risque pour le développement de symptômes de stress post-traumatique au sein des différents groupes de survivants à 5 mois (T1) et 14 mois (T2) de la catastrophe. Une régression hiérarchique a été utilisé afin de déterminer les facteurs de risque significatifs et d'évaluer la proportion de variance qu'ils expliquent. Les symptômes des victimes appartenant aux différentes catégories indiquent bien l'influence du degré d'exposition, de dissociation péri-traumatique et d'insatisfaction vis-à-vis du soutien social reçu sur le développement de symptômes de stress post-traumatique. La dissociation péri-traumatique ressort de l'analyse comme le facteur de risque le plus important par rapport au développement de symptômes de stress post-traumatique à T1. En revanche, à T2, les symptômes de stress post-traumatique présents à T1 sont devenus le plus grand prédicteur de symptômes. L'insatisfaction par rapport au soutien social reçu apparait quant à elle positivement corrélée au développement de symptômes post-traumatiques à T1 et à la présence continuée de ces symptômes à T2. À T1, 6% des répondants montraient une image clinique indicative d'un État de Stress Post-Traumatique (ESPT). À T2, 6,6% des répondants montraient encore une image clinique indiquant un ESPT.

Dans cette suite, le chapitre 7 fournit quant à lui une analyse du développement de quatre troubles de santé mentale dus à l'exposition aux différents aspects de cette catastrophe en comparaison avec des données obtenues dans d'autres études de santé publique dans cette même province. La *Symptom Checklist-90-Revised* (SCL-90-R) a ainsi été utilisée afin de calculer un taux de somatisation, de dépression, d'anxiété et de troubles du sommeil pour trois catégories de victimes impliquées: les témoins directs de souffrance humaine (SHD), les témoins directs n'ayant pas été confrontés à la souffrance humaine (NSHD) et les témoins indirects (IW). Les risques relatifs ont ici été estimés au moyen de modèles de régression logistiques. Les taux de prévalence pour ces quatre troubles de santé mentale étaient plus élevés pour les impliqués appartenant au groupe SHD que pour les autres catégories de victimes, à la fois à T1 et T2. Les groupes NSHD et IW, quant à eux, ne démontraient pas un risque plus élevé de développer un des quatre types de troubles de la santé mentale que les habitants de la même province. Comparée à la population de référence de la même province, à T1 et T2, la catégorie SHD montrait un risque cinq fois plus élevé de développer un trouble de somatisation, quatre fois plus élevé de développer un trouble dépressif ou un trouble du sommeil, et cinq à six fois plus élevé de développer un trouble de l'anxiété. Ceci étant, respectivement à T1 et T2, ils témoignaient qui plus est d'un risque treize et dix-sept fois plus élevé de développer simultanément les quatre troubles de santé mentale précités.

Au chapitre 8, l'analyse est centrée sur les facteurs de risque liés au développement de symptômes de stress post-traumatique chez les enfants impliqués dans la catastrophe. Les réactions des enfants furent également mesurées à 5 mois (T1, N=128) et 14 mois (T2, N=69) après la catastrophe. Respectivement, à T1 et T2, 7% et 4% des enfants répondant montraient des symptômes sévères de stress post-traumatique. Parmi ceux montrant de tels symptômes à T1, 60% en étaient remis à T2. Les facteurs de risque liés au développement de symptômes de stress post-traumatique dans ce cas apparaissent liés au degré d'exposition à la catastrophe, à la dissociation péri-traumatique et à l'insatisfaction vis-à-vis l'aide psychologique reçue.

Enfin, le chapitre 9 discute les conclusions de cette thèse dans le contexte des défis spécifiques posés lorsque qu'une catastrophe technologique frappe soudainement une communauté. Les résultats viennent alors confirmer certains des facteurs de risque bien connus dans le développement et la chronicisation des symptômes de stress post-traumatique, tels que la dissociation péri-traumatique, le degré d'exposition et le manque soutien social ou d'aide psychologique.

En guise d'épilogue, cette thèse propose alors un ensemble de techniques de stabilisation psycho-physiologique innovantes susceptibles de faire toute la différence pour les victimes survivantes d'accidents et de catastrophes pour lesquelles le pire n'est souvent qu'à venir après le crash ou l'explosion.

Dankbetuigingen/Remerciements/ Acknowledgements

De kiemen van deze doctoraatsthesis gaan terug naar het einde van de jaren 90. Ik wil dan ook beginnen met mijn dank uit te drukken naar de militaire collega's die me in die periode hebben aangespoord om zelf het spoor te zoeken naar een meer academische loopbaan. De meeste Europese landen zaten toen volop in het debat rond de effectiviteit van vroegtijdige interventies. Ik verdedigde vroegtijdige interventies zoals *defusing* en *psychological debriefing* met hand en tand op basis van de terreinervaringen die ik opdeed als vrijwillig brandweer-ambulancier en in de schoot van het zelf opgerichte *Fire Stress Team*. Ik zag met lede ogen aan hoe deze interventies als een soort *one-size-fits-all* werden gebruikt voor de opvang van directe slachtoffers in de context van effectiviteitsonderzoek. In de marge van deze discussie, kreeg ik de gelegenheid om vanuit de *Koninklijke Militaire School* een congres te organiseren – in samenwerking met de dienst psychopathologie voor volwassenen van de *Cliniques Universitaires St Luc* – over de opvang na ingrijpende gebeurtenissen.

Een vroege leermeester voor me was Prof. dr. Michel De Clercq van de *Faculté de Médecine* van de *Université Catholique de Louvain*. Michel, je leek als twee druppels water op Luciano Pavarotti en we hebben er vaak om gelachen als mensen je in het buitenland meenden te herkennen. In februari 2000 organiseerden we samen een congres rond psychologische debriefing met sprekers uit Frankrijk, Nederland, Zwitserland en België. We moesten minstens 300 deelnemers hebben om uit de kosten te geraken, maar uiteindelijk tikten we af op meer dan 700 deelnemers. Onmiddellijk in de nasleep van dit congres beslisten we om samen een boek uit te brengen waarin de uiteenlopende standpunten van de auteurs met betrekking tot psychologische debriefing zouden worden opgenomen. Maar... je verdronk in 2001 tijdens een duikvakantie op de Fidji eilanden en ik kwam even alleen te staan. Uiteindelijk heb ik het boek in 2002 samen met mijn collega en vriend Etienne Vermeiren uitgegeven; "*Les débriefings psychologiques en question...?!*" werd een veelgelezen boek en toonde duidelijk de divergentie in standpunten aan tussen de Angelsaksische traumaspecialisten enerzijds, en de Franssprekende psychiaters en psychologen anderzijds. Deze divergentie komt aan bod in deze thesis.

In november 2000 verscheen het eerste nummer van de *Revue Francophone du Stress et du Trauma*, gegroeid uit de werkijver van een handvol Belgische en Franse collega's, met als doelstelling om de Franstalige traumaspecialisten uit landen als Frankrijk, België, Zwitserland en Noord-Afrika een medium te geven om hun theorieën en onderzoek te publiceren in het Frans. Het tijdschrift fungeerde een tiental jaren als een belangrijke referentie in het Franstalige werkveld van de psychotraumatologie. Ik wil Michel, Etienne, Vincent, Louis, François, Alexandre en alle anderen die meewerkten aan de wetenschappelijke vormgeving van de psychotraumatologie in België en Frankrijk bedanken voor deze inspirerende jaren.

In die periode kreeg ik ook de gelegenheid om bestuurslid en ondervoorzitter te worden van de *Association de Langue Française pour l'Etude du Stress et du Traumatisme* (ALFEST). Ik heb deze functie gedurende een tiental jaren op me genomen en wil dan ook graag de mensen die me in de schoot van deze vereniging voorzien hebben van een jarenlange intellectuele voeding, vaak diametraal het tegenovergestelde van wat ik te lezen en te horen kreeg vanuit het Angelsaksische traumaveld, mijn dankbaarheid betuigen.

En première instance, je voudrais témoigner de ma gratitude et mon respect envers le Médecin Général Louis Crocq, Fondateur et Président Honoraire de l'ALFEST, et, Fondateur des Cellules d'Urgence Médico-Psychologique en 1995, pour son soutien, ses leçons et son amitié tout au long de mon parcours d'apprentissage. Grand fut mon plaisir quand l'ISTSS décida de vous attribuer le « ISTSS Lifetime Achievement Award » à Hollywood en 2006. Mes connaissances s'appuient aussi sur les épaules d'autres géants : François Lebigot (avec qui j'ai été jusqu'en Argentine pour exposer les idées francophones en matière de trauma), Guy Briole, Claude Barrois, Carole Damiani, Gilbert Vila, Vincent Dubois, Etienne Vermeiren, Liliane Daligand, François Ducrocq, Guillaume Vaiva, Clara Duchet et tous les autres que j'aurais pu oublier. Vos leçons ont été d'une grande importance pour le développement de mes idées.

La création de l'Association Européenne de Psychologues Sapeurs-Pompiers, en 2003 à l'Ecole Royale Militaire, fut une borne essentielle par rapport au développement de mes connaissances en psychologie de l'urgence. Je dédie la partie de cette thèse en rapport avec les expériences des intervenants, la stabilisation psycho-physiologique et la gestion de catastrophes à ces collègues qui sont pour moi une source d'inspiration et d'amitié profonde : Sylvain Goujard, Valérie Verline, Laurence Auvert, Pascal Perez-Guertault, Alain Zanger, Janny Van Liempd, Marie-Laure Soucarre et tant d'autres, je vous remercie profondément !

Het was Prof. dr. Peter van der Velden die me in de nasleep van een etentje in een Brussels restaurant het advies gaf om bij Prof. dr. Rolf Kleber aan te kloppen om me te begeleiden in een doctoraatsproject. 'Het is een gemotiveerde en strakke begeleider waar je veel zal aan hebben', waren zijn profetische woorden.

Gelijklopend waren er mijn contacten geweest met Prof. dr. Onno van der Hart, in de marge van het *European Conference on Traumatic Stress* in Istanbul (1999), die ik leerde kennen tijdens een pre-conference workshop met Ellert Nijenhuis over structurele trauma-gerelateerde dissociatie van de persoonlijkheid. Ik was sterk onder de indruk van de nieuwe inzichten uit deze workshop.

Kort daarna had ik mijn twee promotoren; Rolf Kleber en Onno van der Hart. Twee klokken van namen, maar van in het begin viel hun beschikbaarheid, hun bescheidenheid en hun toegankelijkheid me op. Ik was op pad vertrokken, als doctorandus, zonder juist te weten welke richting ik zou uitgaan.

Inmiddels had ik ook Prof. dr. Bernard Rimé en Prof. dr. Emmanuelle Zech goed leren kennen in de marge van hun studies over 'social support' en 'social sharing'. Met Emmanuelle Zech begon ik, in samenwerking met het Stressteam van de Federale Politie en een aantal politiekorpsen van de lokale en de federale politie aan een langdurig en intensief studieproject over de opvang van politiepersoneel na kritieke gebeurtenissen.

De doorbraak kwam er op 30 juli 2004, toen ik tijdens mijn verlof abrupt geconfronteerd werd met de ramp die de rechtstreekse aanleiding zou zijn naar het eerste grootschalig rampenonderzoek in België: de gasontploffing in Gellingen (Ghislenghien).

Ik wil de collega's danken die akkoord waren om vrij snel de handen in elkaar te slaan om een breed opgezet en longitudinaal gezondheidsonderzoek te voeren naar de gevolgen van deze gasontploffing. Het werd een samenwerking tussen de Koninklijke Militaire School, het Wetenschappelijk Instituut voor Volksgezondheid en de Université Catholique de Louvain. Ik ben aan deze drie instellingen veel dank verschuldigd.

Aan de leden van het onderzoeksteam en aan de co-auteurs aan de verschillende publicaties hieromtrent wil ik eveneens mijn erkentelijkheid betuigen: Emmanuelle Zech, Ann Versporten, Herman Van Oyen, Michelle Depré, Eva Alisic, Marcia Knarren, Frédéric Daubechies en Sophie Maes. Zonder hen had ik deze thesis nooit kunnen afronden.

Manu, jij krijgt wel een heel speciale bloem toegestuurd voor de vele uren dat je me bijgestaan hebt op statistisch vlak en vooral ook voor de loyaliteit en de vriendschap gedurende de uren waarin we samen op zoek gingen naar betekenissen en verbanden in een uitgebreide dataset.

Na enkele jaren werken kreeg ik vanwege de Academisch Directeur van de Koninklijke Militaire School het verzoek om mijn doctoraatsproject om te vormen tot een co-doctoraat; ik had de eer om ook Prof. dr. Jacques Mylle als promotor in dit onderzoeksproject te betrekken.

Jacques was reeds sedert het najaar 1991 mijn Leerstoelhoofd geweest aan de Leerstoel Psychologie van de Koninklijke Militaire School. Ik leerde van hem om mijn schrijflust in bedwang te houden en alleen nog volgens de '*need to know*' te schrijven. Jaren aan één stuk maakte Jacques tijd vrij om me te helpen met de vormgeving van dit werkstuk. We vormden tevens een goed team aan de Leerstoel Psychologie door de koppeling van experimentele en mathematische kennis aan mijn terreinervaring. Ik vloog in die periode uit over de hele wereld, investeerde veel tijd in het geven en volgen van opleidingen, maar Jacques maande me telkens opnieuw weer aan om aan mijn doctoraatsthesis door te werken. Die boodschap kreeg ik trouwens ook van de opeenvolgende commandanten van de Koninklijke Militaire School. Ik ben hen dankbaar omdat ze de jonge wolf in me steeds hebben aanvaard en tot bedaren trachtten te brengen.

In de totaliteit van dit project ben ik jullie, Rolf, Onno en Jacques, dankbaar voor de wijze waarop ik jarenlang begeleid ben, als buitenbeentje. Jullie moesten erop toezien hoe ik als een bezetene werkte: aanwezig zijn op het terrein van de hulpverlening, bezige bij in elke ramp die België of Belgen in het buitenland trof, deelname aan congressen, boeken schrijven, deelnemen aan militaire zendingen in het buitenland, enz. Maar jullie bleven me geduldig steunen. Rolf, jij bracht een soort empirisch pragmatisme aan de dag en temperde mijn vaak te wilde ideeën.

Onno, jij zat nooit verlegen om een inhoudelijke discussie over trauma en dissociatie. Jacques, jij reed telkens mee op en af tussen Leopoldsburg en Utrecht, zonder morren. Maar telkens vertrok ik weer uit Utrecht met de beide voeten terug stevig op de grond en het gevoel een eeuwige student te blijven. Wijdlopige hoofdstukken werden onder jullie begeleiding strakke empirische stukken. Teruggebracht van 10.000 naar 5.000 woorden. Ook op momenten dat ik me in het wijwatervat roerde als een duivel, bleven jullie me geduldig aansporen om gewoon door te gaan.

Onno, jammer dat ik zo lang treuzelde waardoor je 'jus promovendi' verviel en ik met Rolf en Jacques alleen verder moest.

Ik ben ook dankbaarheid verschuldigd aan de leden van de leescommissie: Prof. dr. Sarah Bal van de Universiteit Gent; Prof. dr. Eric Vermetten van de Universiteit Leiden; en, Prof. dr. Elisabeth Woertman, Prof. dr. Jan van den Bout & Prof. dr. Wilmar Schaufeli van de Universiteit Utrecht. Uw bereidheid om mijn manuscript te lezen tijdens de zomervakantie heb ik bijzonder geapprecieerd.

In de periode van schrijven en veelvuldig vertalen, kon ik ook terugvallen op de hulp van professionele vertalers: Birgit Mylle en Aurélie De Blomme hielpen me hiermee op een erg bijzondere wijze.

Ik wil echter ook Ross McKenzie (Kolonel-Vlieger van de Royal Air Force) en Harold Kudler (US Veterans Administration) danken voor hun waardevolle taalkundige coaching. *Ross, your experiences keep me humble, I thank you for your advice and for sharing your ideas with me. Harold, I am still very grateful for the way in which you reviewed my conceptual chapter on French trauma theory.*

Het heeft me steeds geholpen, al was het maar om er te blijven in geloven, om regelmatig in Utrecht andere promoties te kunnen bijwonen. Ik wil dan ook alle andere promovendi van Rolf, met wie ik altijd goede en vriendschappelijke contacten heb onderhouden, danken voor hun aanmoedigingen en steunbetuigingen.

Het zijn ook mijn functies binnen de Belgische Defensie geweest – respectievelijk als docent aan de Leerstoel Psychologie van de Koninklijke Militaire School en als Commandant van de Sectie Psychosociale Steun van het Stafdepartement Welzijn – die het me mogelijk hebben gemaakt om de laatste vijftien jaar bijna alle congressen bij te wonen van internationale traumaverenigingen (ISTSS, ESTSS, NtVP en zelfs ASTSS). Ik ben hiervoor dan ook erkentelijk naar mijn werkgever toe.

Het Departement Wetenschappelijk en Technologisch Onderzoek bij Defensie van het Koninklijk Hoger Instituut voor Defensie is nu sedert meer dan drie jaar mijn werkplek. Ik dank mijn chefs voor de ruimte en de tijd die ze me gelaten hebben om dit jarenlange project af te werken en voor de toekomstperspectieven die ze me bieden als docent crisispsychologie aan de Koninklijke Militaire School.

The ISTSS Structure & Affiliations Committee, created in 2005, was the first step in an ongoing process on global collaboration and integration with respect for the conceptual and cultural diversities. It has been an honour to co-chair this distinct group. Therefore, I would like to express my explicit gratitude towards Barbara Rothbaum, Paula Schnurr, Danny Kaloupek, Daniel Mosca, Ulrich Schnyder, Merle Friedman, Bessel van der Kolk, Jonathan Bisson and Rick Koepke. I also acknowledge the Board Members of the ISTSS and the ESTSS for providing me the chance to discuss and defend my position on the dissemination of the French-oriented trauma theories within the respective boards and international meetings with the various societies' presidents.

De weg is lang geweest en ik ben onderweg ook veel mensen tegenkomen die me geheel vrijwillig hebben geholpen en me aanspoorden om zeker niet op te geven. De namenlijst zou te indrukwekkend zijn om hier in extenso te weernemen.

Aan het einde van dit dankwoord, volgen zoals gebruikelijk de mensen die me omringen in het dagelijks leven. Mijn lieve vrouw Lies, met wie ik nu vijf jaar samen ben en die me een nieuw geluk heeft geschonken in mijn leven na een onverwachte en pijnlijke echtscheiding. Je hebt me een soort rust en vertrouwen in mijn kunnen gegeven dat ik voorheen niet had. Ik liet een poos de armen hangen en troostte me met het feit dat ik een terreinmens en geen wetenschapper was. Voor jou was het nochtans vanzelfsprekend dat ik dit zou afwerken en met succes verdedigen. Je vond het belangrijk voor mij, net als mijn ex-vrouw Sandra, dat ik dit project niet zou opgeven en je wist daarom meteen de waarde van mijn zwaar skiongeval in te schatten als een unieke kans om deze doctoraatsthesis verder af te werken.

Mijn kinderen Mathias, Steven en Caroline, die het ook altijd belangrijk gevonden hebben dat ik dit zou afwerken. Ik denk dat jullie nu stiekem wel fier zullen zijn, maar ook weten dat je mijn ego er niet te veel mag mee strelen. Mijn pluskinderen, Jasper en Arne, die nog niet weten wat een proefschrift is, maar die het wel 'cool' vinden dat ik ooit 'doctor Ricky' zou genoemd worden.

Mijn naaste familie, mijn moeder op de eerste plaats en mijn vader die dit jammer genoeg niet meer zal meemaken, terwijl hij me altijd in deze richting heeft gestuwd.

Mijn drie zussen, hun partners en kinderen. In het bijzonder mijn zus Ria, die ook mijn meter is en me pardoes een nieuwe laptop kocht – als voorschot op het geschenk dat ik toch zou krijgen voor mijn vijftig jaar – op het moment dat ik met gescheurde kruisbanden vier maanden thuis zat en mijn computer gecrasht was.

Mijn buren, vrienden en collega's.

Tutti gli amici dell'Associazione Psicologi per I Populi in Italia.

All the friends and colleagues from Australia and New Zealand for the way in which you helped me to reach the current level of integration between theory and practice.

Ik ben jullie allen ontzettend dankbaar. Ik ben ontzettend moe na dit parcours en vraag me nog steeds af hoe me dit verder zal helpen op mijn pad.

Maar het is af! Tien boeken verder en meer dan tien jaar later nadat ik het plan opvatte om eens doctoraatsthesis te schrijven.

Ik heb het veel te laat begrepen, maar het wordt mijn nieuw levensdevies: *less is more and the best is the enemy of the good enough!*

Erik DE SOIR

About the author

Erik De Soir (°1965)

Military career

Major Erik De Soir graduated from the Royal Military Academy in 1988 with a Master's Degree in Social & Military Sciences. After fulfilling the Infantry School and the School Company for Para-Commandos, he was assigned to the Liberation Battalion of the 1st Mechanised Brigade as a platoon commander.

In September 1991, he was called back at the Department of Behavioural Sciences of the Royal Military Academy, Chair of Psychology, to teach the courses of general and social psychology. In 1997 he fulfilled several military courses (e.g. the Course for Staff Techniques & the Company Commander's Course) before completing his military training with the Candidate Senior Officer Course at the Royal Higher Institute for Defence in Brussels. In 2009, he was assigned as Commander of the Psychosocial Support Section to the Well-being Department of the Belgian Defence. He is currently the Domain Manager Human Factors & Medicine Research at the Department of Scientific and Technological Research of Defence at the Royal Higher Institute for Defence.

Academic Training

Erik De Soir has a Postgraduate Degree in Disaster Management & Disaster Medicine (Catholic University of Leuven, 1991), a Master's Degree in Clinical Psychology (Catholic University of Leuven, 1995), a Postgraduate Psychotherapy Training in Systemic Marital, Family & Sex Therapy (Catholic University of Leuven, 1998), a Postgraduate Psychotherapy Training in Hypnotherapy (Scientific Flemish Hypnotherapy Association) and extensive training in psychotrauma therapy & counselling (creative arts therapy, EMDR, sensorimotor trauma therapy, cognitive therapy, e.a.).

For several years, he has developed an eclectic model for trauma therapy, based on techniques derived from the traditional Chinese medicine, shiatsu, Buddhist meditation and craniosacral therapy methods.

In 1992, he elaborated the first Belgian Model for Psychosocial Support of Peacekeeping Operations (for soldiers and their significant others) and regularly participated in peace support operations in Somalia, Croatia and Bosnia, to study the different problems of deployed soldiers and their significant others.

He is also both a certified fire-fighter and a paramedic and serves as a volunteer fire psychologist in the Regional Fire Intervention Zone Noord-Limburg (Belgium). He created a European-wide counselling and support network for the management of traumatic stress in fire brigades, emergency medical services and emergency departments of hospitals, currently known under the name of the *Fire Stress Teams* (FiST).

In 2003, he created the *European Association of Fire and Rescue Psychology – Association européenne des psychologie sapeurs-pompiers* (AEPSP) and he is one of the founding members of the *Belgian Society for Psychotraumatology* (*Société belge de psychotraumatologie*) and the *Revue Francophone du Stress et du Trauma.* Between 2001 and 2010, he has been the vice-president of the *Association de langue française pour l'étude du stress et du traumatisme* (ALFEST) and he co-chaired the *International Structure & Affiliations Committee* within the ISTSS (International Society of Traumatic Stress Studies). He is the author of numerous books, book chapters and trauma leaflets, translated in several languages.

Erik De Soir lives in Leopoldsburg, Belgium with Lies Scaut (a psychotrauma therapist, specialized in trauma and grief counselling in children) and together they provide trauma therapy and grief counselling. He is the proud father of three children, Mathias (°1992), Steven (°1994) and Caroline (°1995) and the stepfather of Jasper (°2005) and Arne (°2004).

Publications

Publications related to this thesis

ZECH, E., **DE SOIR, E.**, HANQUET, G., MAES S., VERSPORTEN, A., & VAN OYEN, H. (2005). *Post-traumatic stress disorder after the Ghislengien disaster.* Poster presented at the *Annual Meeting of the European Psychology Association.*

ROOZE, M., NETTEN, J., DE RUYTER, A., DE VRIES, M., HELSLOOT, I., **DE SOIR, E.**, SELWOOD, P., SCHENK, H. **(2008).** Prioritizing Care During the Acute Phase: The Prominent Role of Basic Psychosocial Life Support. *Prehospital and Disaster Medicine.* July-August 2008.

VERSPORTEN, A., **DE SOIR, E.**, ZECH, E., & VAN OYEN, H. (2009). A Longitudinal Study on the Ghislenghien Disaster in Belgium: Strengths and Weaknesses of the Study Design and Influence on Response Rate. *Archives of Public Health.* 67,116-127.

DE SOIR, E., & MYLLE, J. (2010). My Traumatic Journey with Hades: Surviving a Massive Explosion in Ten Progressive Shocks. *Revue Francophone du Stress et du Trauma* – Juin 2010 – Tome 10 – N°2.

DE SOIR, E., KNARREN, M., DEPRE, M., MYLLE, J., KLEBER, R., & VAN DER HART, O. (2011). Expériences potentiellement traumatisantes des secouristes lors d'une catastrophe technologique. *Revue Francophone du Stress et du Trauma. 11(2)*,89-100.

DE SOIR, E., KNARREN, M., DEPRE, M., MYLLE, M., KLEBER, R., & VAN DER HART, O. (2014). Expériences potentiellement traumatisantes des secouristes lors d'une catastrophe technologique. *Journal International de Victimologie. 12(2)*,165-179.

DE SOIR, E., & GOFFINGS, K. (2012). Directe opvang van verkeersslachtoffers. In E. Vermetten, R.J. Kleber& O. van der Hart (Eds.): *Handboek Posttraumatische stressstoornissen.* Utrecht: De Tijdstroom.

DE SOIR, E., KNARREN, M., ZECH, E., MYLLE, J., KLEBER, R., & VAN DER HART, O. (2012). A Phenomenological Analysis of Disaster-Related Experiences in Fire and Emergency Medical Personnel. *Prehospital and Disaster Medicine. 27(2)*,1-8.

DE SOIR, E., ZECH, E., ALISIC, E., VERSPORTEN, A., VAN OYEN, H., KLEBER, R., VAN DER HART, O., & MYLLE, J. (2014). Children Folllowing the Ghislenghien Gas Explosion: PTSD Predictors and Risk Factors. *Journal of Child and Adolescent Trauma.* 7,51-62.

DE SOIR, E., ZECH, E., VERSPORTEN, A., VAN OYEN, H., KLEBER, R., MYLLE, J., & VAN DER HART, O. (2015). Degree of Exposure and Peritraumatic Dissociation as Determinants of PTSD Symptoms in the Aftermath of the Ghislenghien Gas Explosion. *Archives of Public Health, 73:21* (Open Access Online)

DE SOIR, E., VERSPORTEN, A., ZECH, E., VAN OYEN, H., MYLLE, J., KLEBER, R., & VAN DER HART, O. (2015). Does Exposure Time Impact Differentially over Time on the Development of Mental Health Disturbances after a Technological Disaster? *Archives of Public Health, 73,20* (Open Access Online).

DE SOIR, E., KLEBER, R., VAN DER HART, O., & MYLLE, J. (2015). The Concept of Psychological Trauma in Contemporary French Theory: An Unknown Story. (submitted).

Other peer reviewed publications

DE SOIR, E. (1999). *La gestion du stress traumatique chez les sapeurs-pompiers et les ambulanciers. Expériences avec le débriefing psychologique en Belgique,* Médecine de Catastrophe Urgence Collective, 1999 ; 2,139-152.

DE SOIR, E. (2001). *The Management of Emotionally Disturbing Interventions in Rescue Services: Lessons in Humanity, The Quarterly Journal of the Australasian Critical Incident Stress Association,* Vol. 6 : Issue 1, Sep 2001.

DE SOIR, E. (2001). *The Systemic Cost of Deployment: Cohesion in Peacekeepers' Families, The Quarterly Journal of the Australasian Critical Incident Stress Association,* Vol. 6 : Issue 2, Dec 2001.

DE SOIR, E. (2003). Les cultures du trauma, ou le trauma des cultures. *Revue Francophone du Stress et du Trauma. 3(4)*,203-205.

DE SOIR, E. (2005). Normoverschrijdend gedrag tijdens militaire operaties (Hoofdstuk 8). In: *International Humanitair Recht in de kijker – 2005* (Arianne Acke – Ed Rode Kruis Vlaanderen, Brussel.

GRAUX, C., **DE SOIR, E.** (2005). Visites à domicile au décours d'une inondation. Des psychologues "hors murs" à pratique "hors normes". *Revue Francophone du Stress et du Trauma. 5(2)*,109-115.

DE SOIR, E. (2005). El manejo de intervenciones emocionalmente perturbadoras durante incendios de rescate. *Revista de Psicotrauma para Iberoamerica. IV(1)*,4-20.

DE SOIR, E. (2005). Las cultural del trauma... o el trauma de la culturas. *Revista de Psicotrauma para Iberoamerica. IV(2)*,39-40.

NIJENHUIS, E., VAN DER HART, O., STEELE, K., **DE SOIR, E.**, & MATTHESS, H. (2006). Dissociation structurelle de la personnalité et trauma. *Revue Francophone du Stress et du Trauma. 6(3)*,125-139.

DE SOIR, E. (2006). Pr Louis Crocq. Lifetime Achievement Award 2006. Revue Francophone du Stress et du Trauma. *Revue Francophone du Stress et du Trauma. 6(4)*,195-196.

VAN DER HART, O., STEELE, K., NIJENHUIS, E., & **DE SOIR, E.** (2009). Souvenirs traumatiques. Leur traitement selon le modèle de la dissociation structurelle de la personnalité. *Revue Francophone du Stress et du Trauma. 9(2)*,81-92.

DE SOIR, E. (2013). Crisispsychologie en psychotraumatologie in Vlaanderen: een uitdaging voor psychologen. *Tijdschrift voor Klinische Psychologie. 43*,287-300.

Publications – books

DE SOIR, E. (1997). *Traumatische stress en politie.* Antwerpen/Apeldoorn: Maklu Uitgevers.

DE SOIR, E. (2000). *Op het netvlies gebrand ...! Traumatische stress bij hulpverleners.* Antwerpen/Apeldoorn: Garant Uitgevers.

DE SOIR, E., & VERMEIREN, E. (2002). *Les débriefings psychologiques en question ... !?* Antwerpen/Apeldoorn: Garant Uitgevers.

DE SOIR, E. (2004). *Marqué au cœur ... Le trauma des équipes de secours.* Antwerpen/ Apeldoorn : Garant Uitgevers.

DE SOIR, E. (2006). *Een heel klein beetje oorlog... Omgaan met traumatische ervaringen.* Tielt: Lannoo.

DE SOIR, E., DAUBECHIES, F., VAN DEN STEENE, P. (2007). *Stress en trauma bij de politie.* Antwerpen/Apeldoorn: Maklu Uitgevers.

DE SOIR, E., DAUBECHIES, F., VAN DEN STEENE, P. (2007). *Stress et trauma en milieu policier.* Antwerpen/Apeldoorn: Maklu Uitgevers.

DE SOIR, E., DAUBECHIES, F., VAN DEN STEENE P. (2012). *Stress et trauma dans les services de police et de secours.* MAKLU Uitgevers.

DE SOIR, E. (2013). *Redders in nood. Opvang van mensen in crisis.* Tielt: LannooCampus.

DE SOIR, E. CELLI, TIZIANA. (2015). *Colipiti al cuore. Il trauma nelle squadrea di soccorso* (in press)

DESOIR, E. (2015). *Marcado en el corazon* (in press).

DE SOIR, E. (2015). *Leven in de zijlijn... Zingeving bij trauma en verlies.* Tielt: Lannoo (in press).

DE SOIR, E., DAUBECHIES, F., & VAN DEN STEENE, P. (2015). *Stress en trauma bij de politie- en de hulpdiensten.* MAKLU Uitgevers.

Book chapters

DE SOIR, E. (2000). *Aufbau von krisenpsychologischer Hilfeleistung und Debriefing in der Feuerwehr und im ärztlichen Rettungsdienst.* In: G. PERREN-KLINGLER (Ed.). Debriefing – *Erste Hilfe durch das Wort Hintergründe und Praxisbeispiele,* Bern-Stuttgart-Wien: Verlag Paul Haupt.

DE SOIR, E. (2000). La gestion du stress traumatique chez les sapeurs-pompiers et les ambulanciers. In: G. PERREN-KLINGLER (Ed.). Maux en mots. Debriefing: modèles et pratiques. Bern-Stuttgart-Wien: Verlag Paul Haupt.

DE SOIR, E. (2000). La gestion del estrès traumatico en bomberos y conductores de ambulancia. In G. PERREN-KLINGLER (Ed.). Debriefing: Modelos y Aplicationes. De la historia traumatica al relato integrado.

DE SOIR, E. (2000). *Les intervenants des services de secours lors de situations d'exception : les bases de la traumatisation indirecte.* In : Les Traumatismes Psychiques (DE CLERCQ, M. & LEBIGOT, F. ; Eds.), Masson : Paris.

DE SOIR, E. (2006). Psychosocial Crisis Intervention with Military and Emergency Services Personnel. In : G. Reyes and G.A. Jacobs, Handbook of International Disaster Psychology (Vol.4). Westport, Connecticut/London : Praeger Perspectives.

DE SOIR, E. (2012). The Management of Emotionally Disturbing Interventions in Fire and Rescue Services :Psychological Triage as a Framework for Acute Support. In : R. Hughes, A. Kindler & C.L. Cooper, *International Handbook of Workplace Trauma Support*. West Sussex : John Wiley & Sons.

DE SOIR, E. (2015). Psychosociale interventies voor first responders bij calamiteiten en rampen. In: E. Devroe, A. Duchatelet, M. Easton, L.G. Moor, L. Wontergem & P. Ponsaers (Eds.), *Zicht op first responders. Handboek bij het beheer van evenementen en noodsituaties in Nederland en België (pp.281-296)*. Antwerpen/Apeldoorn: Maklu Uitgevers (in press).

DE SOIR, E. (2015). Psychological Support After Long Term Deployment: From Third Location Decompression to Readjustment. In S. Bowles & P.Bartone (Eds.), *The Practice of Military Psychology*, New York: Routledge (in press).

Publications in fire, rescue & crisismanagement magazines

DE SOIR, E. & SERNICLAES, O. (1995). *Maîtrise et prévention des* événements traumatiques, Urgence Pratique du médecin généraliste au spécialiste : Information Santé, Incendie & Secours. October 1995;13,25-29.

DE SOIR, E. (1996). *Le stress chez les pompiers. Accidents de la circulation avec personnes incarcérées : les conséquences psychologiques. Tu as eu beaucoup de chance, tu as eu plus de peur que de mal ... !* Le Sapeur-Pompier belge. 4,21-25.

DE SOIR, E. (1997). *Preventie, aanpak en nazorg van traumatische stress: de Fire-Fighter Stress Teams,* De West-Vlaamse Brandweerman,273,6-8

DE SOIR, E. (1997). *Accidents de la circulation avec personnes incarcérées: les conséquences psychologiques.* Urgence pratique du médecin généraliste au spécialiste : Information Santé, Incendie & Secours, 22,15-19.

DE SOIR, E. (1997). *Traumatische stress : preventie, aanpak en nazorg bij de EHBO-er,* NVBB Magazine,137,38-45.

DE SOIR, E. (1997). *Stress traumatique: prévention, prise en charge et suivi chez les secouristes,* ANPI-Magazine,137,38-45.

DE SOIR, E. (1997). *Prévention, prise en charge et suivi du stress traumatique chez les secouristes : les Fire-Fighter Stress Teams seront-ils les pilotes ?,* Urgence Pratique du médecin généraliste au spécialiste : Information Santé, Incendie & Secours,25,5-9.

DE SOIR, E. (1997). *Accidents de la circulation avec personnes incarcérées : les conséquences psychologiques,* Journal des sapeurs-pompiers suisses,6,500-504

DE SOIR, E. (1997). *Prévention et gestion du stress traumatique.* Journal des sapeurs-pompiers suisses, 6,505-507

DE SOIR, E. (1997). *Les Fire-Fighter Stress Teams seront-ils les pilotes ? Prévention, prise en charge et suivi,* Journal des sapeurs-pompiers suisses, 9,734-739.

DE SOIR, E. (1998). *Les Fire-Fighter & Medical Emergency Stress Teams,* Urgence Pratique du médecin généraliste au spécialiste: Information Santé, Incendie & Secours, 26,5-6.

DE SOIR, E. (1998). *Call for European-wide crash trauma counselling,* Fire International, 164,21-23.

DE SOIR, E. (1998). *Need for European-trauma counselling,* Fire Europe, 7,7-9.

DE SOIR, E. (1998). *Pompierii çi stresul profesional,* Pompierii Romana – Revista ilustrata de analiza si sintezo, cultura tehnica si informare publica anti incendiu, 11,28-29

DE SOIR, E. & STROBL, D. (2002). *Rodiny vojaku mirovych misi* – Psychologie Dnes, 3, 28-29.

DE SOIR, E. & STROBL, D. (2002). *Navrat domu neni snadny* – Psychologie Dnes, 4,26-27.

DE SOIR, E. (2004). *Prise en charge psychologique après l'enfer de Ghislengien. Accompagnement psychologique des intervenants,* ANPI Magazine, 172.

DE SOIR, E. (2004). *Psychologische opvang na de hel van Gellingen. Psychologische begeleiding van de hulpverleners.* NVBB Magazine,172.

DE SOIR, E. (2005). *Psychosociale aspecten – Psychosociaal verliesbeheer.* Nieuwsbrief KLUWER, Week 50 – N° 23 Jaargang 24.

DE SOIR, E. (2006). Communicatie – Leiderschap in crisisomstandigheden. Nieuwsbrief KLUWER; Week 3 – N° 2 Jaargang 25.

DE SOIR, E. (2007-2008). Psychosocial crisis intervention – Part 1 – 4. Vol 2 – Issue 4; Vol 3 – Issues 1 – 2 – 3 – 4; Vol 4 – Issue 1. In: CRISIS REPONSE JOURNAL.

SPAEPEN, K., **DE SOIR, E.** (2007). Acute psychologische opvang van kinderen na een traumatogene gebeurtenis: fictie of feit? In: *Spoedgevallen* – Jaargang 26 - Nr 4 – September - October.

SPAEPEN, K., **DE SOIR, E.** (2008). *Psychological first aid for children after a traumatic experience.* CRISIS RESPONSE JOURNAL, Vol. 4, Issue 3.

DE SOIR, E. (2012). Uitbouw van een psychosociaal traject voor nazorg bij collectieve urgenties op basis van psychologische triage. *PREBES Veiligheidsnieuws* (Nr 176), 46ste jaargang.

DE SOIR, E. (2012). *Dealing with school tragedy.* CRISIS RESPONSE JOURNAL, Vol. 8, Issue 2.

DE SOIR, E. (2012). Het abc van psychotraumatherapie. Van speechless terror tot integratie in het levensverhaal. *Psyche.* Jaargang 24(4).

DE SOIR, E. (2012). Uitbouw van een psychosociaal traject voor nazorg bij collectieve urgenties op basis van psychologische triage. Prebes Veiligheidsnieuws. 174,40-43.